Project Management for Engineering Design

Project Management for Engineering Design
Charles Lessard and Joseph Lessard
www.morganclaypool.com

ISBN-10: 1598291742   paperback
ISBN-13: 9781598291742   paperback

ISBN-10: 1598291750   ebook
ISBN-13: 9781598291759   ebook

DOI 10.2200/S00075ED1V01Y200612ENG002

A Publication in the Morgan & Claypool Publishers Series
SYNTHESIS LECTURES ON ENGINEERING #2
Lecture #2

Series ISSN: 1559-811X   print
Series ISSN: 1559-8128   electronic

First Edition
10 9 8 7 6 5 4 3 2 1

Printed in the United States of America

Project Management for Engineering Design

Charles Lessard
Texas A&M University

Joseph Lessard
Globeleq Inc.

SYNTHESIS LECTURES ON ENGINEERING #2

MORGAN & CLAYPOOL PUBLISHERS

ABSTRACT

This lecture book is an introduction to project management. It will be of use for engineering students working on project design in all engineering disciplines and will also be of high value to practicing engineers in the work force. Few engineering programs prepare students in methods of project design and configuration management used within industry and government.

This book emphasizes *teams* throughout and includes coverage of an introduction to project management, project definition, researching intellectual property (patent search), project scope, idealizing & conceptualizing a design, converting product requirements to engineering specifications, project integration, project communications management, and conducting design reviews. The overall objectives of the book are for the readers to understand and manage their project by employing the ***good engineering practice*** used by medical and other industries in design and development of medical devices, engineered products and systems. The goal is for the engineer and student to work well on large projects requiring a *team* environment, and to effectively communicate technical matters in both written documents and oral presentations.

KEYWORDS

Engineering Design, Teams, Conflict Resolution, Decision Making, Project Management, Time Management, Cost Management, Risk Management, Earned Value Analysis.

Contents

CHAPTER 1

Introduction to Engineering Design

The engineering design process is similar to problem-solving processes taught in engineering colleges. Most commercial products begin by identifying the commercial market needs of some novel "ideas," which may impact some commercial market. Hence, the first steps in a product design may include

1. novel idea, or potential market needs,

2. forming a team, since most complex problems or products are not developed by a single individual,

3. development of generalized broad product requirements:
   a. Generally, user requirements may not be technical; e.g., a doctor wants to monitor temperature, or a veterinarian wants to remove purring sound from a stethoscope so he can hear the cat's heart valve sounds. What are the technical specifications?

4. define required input and outputs,

5. define the product outcome:
   a. What is it used for, by whom? Inputs?
   b. How do we measure the desired outcome?
   c. How do we test the desired outcome?
   d. What are the criteria for acceptance?

1.1   TEAMS

Webster [1] defines the word "team—a noun" in various ways:

1. Two or more draft animals harnessed to a vehicle or farm implement.

2. A vehicle along with the animal or animals harnessed to it.

3. A group of animals exhibited or performing together.

4. A group of players on the same side in a game.

5. Any group organized to work together.

Engineering has expanded the definition to "A team is a *small group of people* with *complementary skills* who are committed to a *common purpose, performance goals,* and approach for which they hold themselves *mutually accountable*" [2].

You might wonder why work in teams? One might cite that "Team skills are valued by industry" or that one person does not have in-depth knowledge in all the disciplines and functions of engineering to design and produce a spacecraft or a space station. It requires a "team" of people with diverse backgrounds in the necessary technical areas to produce the product for space. Simply put, teams need engineers with a broad set of talents, i.e.

1. technical knowledge
2. creativity
3. people skills
4. planning ability
5. management skills.

Diverse abilities and diverse ways of thinking and viewing problems is the strength of teams that function well together. Because teams are important in the success of projects, project managers must understand teams and consider five issues in team building.

1. *Interdependence*: It is the issue of how each member's outcomes are determined, at least in part, by the actions of the other members. Functioning independently of one another or competing with your teammates may lead to suboptimal or disastrous outcomes for both the entire team and the project.

2. *Goal specification*: It is very important for team members to have common goals for team achievement, as well as to communicate clearly individual goals that members may have.

3. *Cohesiveness*: It refers to the attractiveness of the team membership. Teams are cohesive to the extent that membership in them is positively valued, that is, members are drawn toward the team. Patterns of interpersonal attraction within a team are a very prominent concern. Task cohesiveness refers to the way in which skills and abilities of the team members mesh to allow effective performance.

4. *Roles and norms*: All teams need to develop a set of roles and norms over time.
   a. *Roles*: For a student team, the role structure will enable the team to cope more effectively with the requirements of a given task. The roles may be rotated so that all team members experience, and learn from, the various positions held. It is extremely important that the roles are understood and accepted by team members.

b. *Norms*: For a student team, norms are the rules governing the behavior of team members, and include the rewards for behaving in accord with normative requirements, as well as the sanctions for norm violations. It is not uncommon for a set of norms to develop between team members that are never actively discussed. However, it is always better to have interaction rules appear in the form of a written document, such as in a "code of cooperation": the agreed upon rules governing the behavior of team members, as well as any appropriate rewards and sanctions. The team code of cooperation sets a norm for acceptable behavior for each team member and represents how the team members will interact with one another. It should be developed, adopted, improved, and/or modified by all team members on a continuous basis; and copies should be easily accessible by team members.

5. *Communication*: Effective interpersonal communication is vital for the smooth functioning of any task team. It is also important for a team to develop an effective communication network: who communicates to whom; is there anybody "out of the loop?" Norms will develop by governing communication. Do those norms encourage everyone to participate, or do they allow one or two dominant members to claim all the "air time?" [3].

Key team roles include the following:

1. *Meeting coordinator*: Coordinates and prepares the agenda (i.e., what needs to be accomplished, establishes a process, etc.); coordinates time, date, and place of meetings; ensures all necessary resources are available for the meetings; keeper of the code of cooperation (to be discussed); monitors the decision-making process; coordinates the process check. However, this person is *not* the "boss."

2. *Recorder*: The recorder is the person responsible for doing the writing of the team whenever group work is being done, which should maximize participation by the rest of the team, since no one else needs to worry about it. If required, the recorder also ensures that the process(es) being used by the team is (are) documented and/or prepares an "action list" to keep a record of the assigned actions. In addition, the recorder makes sure that copies of their work are provided to the rest of the team.

3. *Time keeper*: The time keeper has the responsibility of keeping tract of time, as well as keeping the team moving so that they can finish the task at hand.

4. *Encourager/gatekeeper*: The encourager/gatekeeper has the task of giving encouragement to all the other team members. The person also has the responsibility of maintaining a balanced level of participation for all the members. They will encourage the silent members and try to hold back the verbose, dominate members.

5. *Devil's advocate*: The devil's advocate takes a position opposite to that held by the team to ensure that all sides of an issue are considered. This responsibility should be undertaken by all team members [4].

In a learning environment such as student teams, the roles should rotate among team members. In summary, effective teams include the use of roles, the development of a code of cooperation, the use of the check for understanding to make sure everybody is "on the same page," the development of effective listening skills, the ability to give and take effective constructive feedback, the use of agendas for planning meetings, the use of contact before work to provide time for nontask-related discussions, the definition of decision-making processes to be included in the agenda, the use of the issue bin to provide time for discussion of items not in the agenda, the use of *action* lists to keep a record of assigned actions, the use of a process check for continuous improvement, and a commitment from *all* the members of the team. Once the team is established, the purpose and objectives of the team should be defined and documented. Subsequently, the project must be adequately defined.

1.2   DEFINING THE PROJECT OR PROBLEM

One of the most important aspects in product development and engineering design is to adequately define the scope of the problem. Often, the problem is stated initially in terms of vague project requirements. The team must redefine the product requirements in terms of inputs, output, and appearances, then convert and link requirements to "technical specifications," e.g., performance, accuracy, tolerances, etc. One should keep in mind that "all specifications *must* be tested." Additionally, the team must develop and document "pass or fail acceptance criteria" for each specification, as well as goals or criteria for success and constraints (part of scope). Typical goals or criteria for success include aesthetics, performance, quality, human factors, costs ("initial capital" and "life cycle" costs), safety, operating environment, interface with other systems, effects on surroundings, logistics, reliability, maintainability (preventive and corrective maintenance), serviceability, and availability. Constraints usually include the following factors: budget, time, personnel, legal, material properties, availability of materials, off-the-shelf purchase versus fabrication/construction, competition, and manufacturability (can it be manufactured?) [5].

1.3   BACKGROUND

Investors or a company marketing survey may have expressed the need for a new or a better product as an idea or in very general terms, but engineers working on the project design require greater background knowledge. Knowledge in the form of publications on other similar designs that may be found in

1. library literature searches,
2. Web-based searches, or
3. patent searches.

Patent searches are a key element in products being designed and developed for the commercial sector, since patent infringement may lead to lawsuits. Without doubt, searches take a lot of time, and results of the searches must be analyzed, documented, and reported in the "background" section of the proposal, such as a small business incentive research (SBIR) proposal. This step also takes a lot of time, often months; it may be continued throughout the product development period.

1.4   DESIGN PHASES

Most student teams believe design is a single-step process, which it is not. Designing in industry is a multipass process that includes various phases: the *conceptualization phase*, *feasibility study phase*, *preliminary design phase*, and *detailed design phase*. In the feasibility study phase, conceptualization and "brainstorming" ideas are roughed out.

In searching for a solution, selection of the most feasible ideas for refinement is a team decision. The following heuristics are often applied in the search for a solution:

1. challenge basic assumptions
2. employ analogies
3. identify critical parameters
4. switch functions
5. alter sequence of steps
6. reverse the problem
7. separate or combine functions
8. use vision
9. employ basic engineering principles.

In the preliminary design phase, the most promising ideas are explored and analyzed in more detail. Finally, in the detailed prototype design phase, the team develops highly detailed

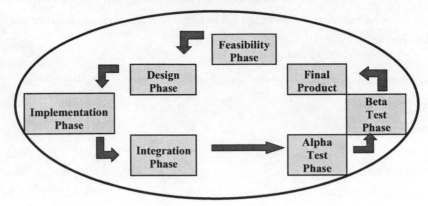

FIGURE 1.1: The design methodology of IWT on "real-world" experience

drawings with final specifications prepared for the "best design option." In all design phases, the following four steps are repeated:

1.   analyzing each potential solution,
2.   choosing the best results or solutions,
3.   documenting the results or solutions,
4.   communicating results or solutions to management.

A detailed design phase is when companies usually construct or fabricate prototypes. Marketing strategies would be in development in parallel about this time. Once the prototype is finished, it undergoes "verification" or "alpha" testing and evaluation of design specifications, performance parameters, safety, etc. The results are compared against the acceptance criteria. If the specification fails the acceptance criteria, the prototype may undergo a redesign. If all tests are acceptable, the product may undergo validation or "beta" testing, which means the product will be sent to various users to test the product with noncompany subjects.

Innovative Wireless Technologies (IWT) structured their strict design methodology on "real-world" experience (Fig. 1.1). Their feasibility phase includes requirements, project budget, project schedule, and design specifications. In their design phase, IWT includes technical patent search, functional design specs (FDS), FDS design review, and product verification plan. The implementation phase includes prototype development, integration test plan, patent review, factory prototype review, and product launch. The integration phase includes plans and documentation for factory prototype development, integration testing, and environmental testing. In the alpha test phase (verification), test engineers conduct alpha testing, preseries production, release the design, and develop the beta test plan. Validation test phase includes beta testing, release customer documentation, initial series production, yield analysis, and training [6].

Testing in industry is not the same as a test conducted by a student in academic laboratories. The team must develop test plans and test procedures before the breadboard and prototype system is developed and tested. In the verification testing, the prototype (hardware and software) is also verified in an integrated test environment with all necessary test equipment for EMI/EMC and pre-FCC verification.

1.5   SUMMARY

Project managers and their teams should focus their attention and efforts on meeting project objectives and producing positive results. It is recommended that instead of blaming team members, the focus should be turned to fixing the problem. Project managers should establish regular, effective meetings with agenda and openness. Part of the program managers tasks include nurturing team members, encouraging them to help each other, and to acknowledge in public the individual and group accomplishments. Additional information on teams may be found in [7, 8].

REFERENCES

[1]   *Webster's New Collegiate Dictionary*, G&C Merriam Co., Springfield, MA, 1973.

[2]   J. R. Katzenbach and D. K. Smith, *The Wisdom of Teams*. Boston, MA: Harvard Business School Press, 1993.

[3]   *Surviving the Group Project: A Note on Working in Teams* [Online]. Available: http://web.cba.neu.edu/~ewertheim/teams/ovrvw2.htm, 2005.

[4]   D. A. Nadler, *Designing Effective Work Teams*. New York: Delta Consulting Group, 1985.

[5]   G. P. Shea and R. A. Guzzo, "Group effectiveness: What really matters," Sloan Manage. Rev., vol. 3, pp. 25–31, 1987.

[6]   *Innovative Wireless Technologies (IWT) Design Methodology* [Online]. Available: http://www.iwtwireless.com, 2004.

[7]   V. R. Johnson. (2005, Mar.). Understanding and assessing team dynamics. *IEEE-USA Today's Eng.* [Online]. Available: http://www.todaysengineer.org/2005/Mar/team_dynamics.asp.

[8]   V. R. Johnson. (2004, Nov.). Understanding and assessing team dynamics. *IEEE-USA Today's Eng.* [Online]. Available: http://www.todaysengineer.org/2004/Nov/self-assessment.asp.

CHAPTER 2

Project Management Overview

Before diving into project management, let us begin with a simple question. What is a "project"? A project is a temporary endeavor undertaken to accomplish a unique purpose. I have managed projects ranging from simple projects, like the development of an automated EEG analyzer, to complex high dollar value, such as the installation of Spain's Air Defense System. Yet, regardless of complexity, all projects have similar attributes:

1. Each project has its own "unique purpose."
2. Projects are "temporary" with time constraints.
3. Projects require resources (manpower, funding, and materials), often from various areas.
4. Commercial projects should, and usually, have a primary sponsor and/or customer.
5. All projects involve uncertainty.
6. Every project is constrained in different ways by its
   a. scope goals
   b. time goals
   c. cost goals.

It is the project manager's responsibility to balance these three competing goals. So, what is project management? Project management is "The application of knowledge, skills, tools, and techniques to project activities in order to meet or exceed stakeholder needs and expectations from a project" [1].

In the definition of project management, the terminology is often misinterpreted to mean investors with stock in the company or "stockholders"; whereas, "stakeholders" are the people involved in or affected by project activities. Thus, "stakeholders" include the project sponsor and all members of the project team, support staff, customers, users, suppliers, and even opponents to the project.

2.1   PROJECT MANAGEMENT KNOWLEDGE AREAS

Knowledge areas describe the nine key competencies that project managers must develop. There are four core knowledge areas that lead to specific project objectives (scope, time, cost, and quality). There are also four facilitating knowledge areas that are the means through which the project objectives are achieved (human resources, communication management, risk management, and procurement management). The final knowledge area (project integration management) affects and is affected by all of the other eight knowledge areas. Although much of the knowledge needed to manage the projects is unique to project management, nevertheless, project managers must also have knowledge and experience in "general management" and in the application area of the project. Ultimately, project managers must focus on meeting specific project objectives. This book will develop and elaborate on each of the nine knowledge areas in separate chapters.

There are several project management tools and techniques that assist project managers and their teams in various aspects of project management. Some specific tools include

1. project charter
2. Work breakdown schedule (WBS) or scope
3. Gantt charts, PERT charts, critical path analysis (time)
4. Cost estimates and earned value analysis (cost).

Most of these tools are developed with software programs, such as Microsoft Project 2003.

So what is the advantage of implementing project management on any project? The advantages that program management offers might include that "Bosses," customers, and other stakeholders do not like surprises especially, "Bad News Surprises." Good project management provides assurance and reduces the risk of project failure or large cost overrun. Project management provides the tools and environment to plan, monitor, track, and manage schedules, resources, costs, and quality of the product (project). Project management also provides a history or metrics base for future planning as well as good documentation, which is required by the Food and Drug Administration (FDA) and Good Manufacturing Practice. Perhaps for the students, the greatest advantage is that project team members learn and grow by working in a cross-functional team environment.

Some books contend that "modern project management" began with the Manhattan Project, which the U.S. military led to develop the atomic bomb. Yet, some may argue that it was not until systems approach emerged in the 1950s which described a more analytical approach to management and problem solving that modern project management really began. The systems approach to project management includes three parts:

1. *Systems philosophy*: Project managers should view projects and things as systems, interacting components working within an environment to fulfill some purpose.

2. *Systems analysis*: Project managers should use a problem-solving approach, which engineering students are taught.

3. *Systems management*: Project managers should address business, technological, and organizational issues before making changes to systems.

Project managers need to take a holistic or systems view of a project and understand how it is situated within the larger organization, since projects developed must operate in a broad organizational environment; meaning, "projects cannot be run in isolation."

2.2   PROJECT LIFE CYCLES AND PROJECT PHASES

A project life cycle is a collection of project phases, which vary with the project or industry. Table 2.1 shows some general phases that include concept, development, implementation, and support.

2.3   PRODUCT LIFE CYCLES

Products also have life cycles. The systems development life cycle (SDLC) is a framework for describing the phases involved in developing and maintaining information systems. Typical SDLC phases include planning, analysis, design, implementation, and support. There are

TABLE 2.1: Phases of the Project Life Cycle

PROJECT FEASIBILITY		PROJECT ACQUISITION	
Concept	Development	Implementation	Closeout
Management plan	Project plan	Last work package	Completed work
Preliminary cost estimates	Budgetary cost estimates	Definitive cost estimates	Lessons learned
3-level WBS	6+—level WBS	Bulk of time spent in this phase	Customer acceptance

several SDLC models such as

1. the waterfall model that has well-defined, linear stages of systems development and support,

2. the spiral model which shows that products are developed using an iterative approach rather than a linear approach.

In addition, there are the incremental release model and the prototyping model that are used for developing prototypes to clarify the user requirements.

Project life cycle applies to all projects, regardless of the products being produced, and product life cycle models vary considerably based on the nature of the product. Most large projects are developed as a series of smaller projects, and then integrated. Project management activities are done through the entire product life cycle phases. A project should successfully pass through each of the project phases in order to continue on to the next phase of the life cycle. To verify that all the requirements of a phase were completed satisfactory, the program manager should conduct project reviews (also called project management review or program management review) at preset project milestones. Management reviews (often called phase exits or kill points) should occur after each phase to evaluate the project's progress, likely success, and continued compatibility with organizational goals.

2.4   ORGANIZATIONAL STRUCTURES

To understand how the various organizational structures and frames can help or impede the program manager in product development, one needs to understand organizations. There are four basic organizational frames:

1. The structural frame that focuses on roles and responsibilities, coordination and control. Organization charts help define this frame.

2. The political frame that assumes organizations are coalitions composed of varied individuals and interest groups. Conflict and power are key issues within this frame.

3. The human resources frame that focuses on providing harmony between needs of the organization and needs of the people.

4. The symbolic frame that focuses on symbols and meanings related to events. In this frame, culture is important.

Most managers and people understand what organizational charts are; yet, many new managers try to change organizational structure rather than concentrating on other changes that are really needed. There are three basic organization structures: functional, project, and matrix, as shown in Table 2.2. The first column in Table 2.2 lists project characteristics, and their influence on the project based on the type of organizational structure are compared in the rows. The table also indicates that the project-oriented organizational structure provides the project manager with the highest level of authority, personnel and administrative staff that are assign "full-time" to work on the project, and that his role and title are "full-time" and "project manager," respectively. Although the organizational structure influences the project manager's authority, project managers also need to remember and address the human resources, political, and symbolic frames.

Recall that project stakeholders are the people involved in or affected by project activities; hence, project managers must take time to identify, understand, and manage relationships with all project stakeholders. Using the four frames of organizations can help meet stakeholder needs and expectations.

2.5   PROJECT MANAGEMENT JOB FUNCTIONS

At this point, you (the reader) may still be asking, "But what does the project manager do?" Most organizations establish job positions with a description of the responsibilities and functions of the position. The job description for the position of project manager usually requires that the project manager define the scope of project, form a team, identify stakeholders, identify decision-makers, and establish escalation procedures should the project encounter major problem requiring a higher level decision. He is also responsible for the development of a detailed task list or work breakdown structures for the project. Additionally, he is responsible for the estimation of the time requirements not only for the project, but also for each task in the work breakdown schedule. The project manager is responsible for the development of initial project management flow chart and identification of required resources with budget estimates. He evaluates the project requirements, identifies and evaluates the risks, and is responsible for preparing contingency plans.

The program manager must identify interdependencies within and outside of the organization. He is required to identify and track critical milestones, and conduct or participate in project progress and phase reviews. He has the responsibility of securing the needed resources in a timely manner. Additionally, the program manager is responsible for the management of the change control process, which may require establishment of a "change control board" to administer the handling of product configuration and changes to the baseline configuration. The final job function of project managers is the collection of information and preparation of project status reports in documents and in presentation at higher level "program reviews" [2].

TABLE 2.2: Influences of Organization Structure on Projects

| PROJECT CHARACTERISTICS | FUNCTIONAL | ORGANIZATIONAL TYPE | | | PROJECTIZED |
| | | MATRIX | | | |
		WEAK MATRIX	BALANCED MATRIX	STRONG MATRIX	
Project manager's authority	Little or none	Limited	Low to moderate	Moderate to high	High to almost total
Performing organization's personnel assigned full-time to project work	Virtually none	0–25%	15–60%	50–95%	85–100%
Project manager's role	Part-time	Part-time	Full-time	Full-time	Full-time
Common title for project manager's role	Project coordinator, project leader	Project coordinator, project leader	Project manager, project leader	Project manager, program manager	Project manager, program manager
Project management administrative staff	Part-time	Part-time	Part-time	Full-time	Full-time

TABLE 2.3: Comparison of Characteristics of Effective and Ineffective Project Managers

EFFECTIVE PROJECT MANAGER	INEFFECTIVE PROJECT MANAGER
Leadership by example	Sets bad example
Visionary	Not self-assured
Technically competent	Lacks technical expertise
Decisive	Indecisive
Good communicator	Poor communicator
Good motivator	Poor motivator
Stands up to upper management when necessary	
Supports team members	
Encourages new ideas	

It is strongly suggested by the author that the project managers develop the following skills:

1. *Communication skills*: listening and persuading. From elementary grades, we were taught how to speak, read, and write, but we were not taught how to listen. Communication theory requires three elements for effective communication: a transmitter (the speaker), a common media (the language), and the receiver (the *listener*).
2. *Organizational skills*: planning, goal setting, and analyzing.
3. *Team building skills*: people skills, empathy, motivation, and esprit de corps.
4. *Leadership skills*: sets example, energetic, vision (big picture), delegates, and positive.
5. *Coping skills*: flexibility, creativity, patience, and persistence.
6. *Technological skills*: technical knowledge, project knowledge, and experience.

Table 2.3 compares the most common characteristics found amongst "effective" and "ineffective" project managers.

In summary, project management may be viewed as a number of interlinked management processes that include initiating processes, planning processes, executing processes, controlling processes, and closing processes. Table 2.4 shows the relationship between the project manager's knowledge areas, project processes, and the activities required in project management. From

TABLE 2.4: Relationships Among Project Processes, Activities, and Knowledge Areas

KNOWLEDGE AREA	INITIAL-IZING	PLANNING	EXECUTING	CONTROLLING	CLOSING
Integration		Project plan development	Project plan execution	Overall change control	
Scope	Initiation	Scope planning	Scope verification	Scope change control	
Time		Activity definition, sequencing, duration estimation			
		Schedule development			
Cost		Resource planning		Cost control	
		Cost estimating, budgeting			
Quality		Quality planning	Quality assurance	Quality control	
Human resources		Organizational planning	Team development		
		Staff acquisition			
Communications		Communication planning	Information distribution	Performance reporting	Administration closure
Risk		Risk identification, quantification, response development	control	Risk response	
Procurement		Procurement planning	Solicitation, source selection		Contract closeout
		Solicitation planning	Contract administration		

the table, one can surmise that the project manager will use all knowledge areas in the planning phase, but must apply integration, scope, quality, communications, and procurement knowledge areas throughout the project processes or "life cycle" [3]. On the other hand, team members will spend the majority of their time in the "execution" phase of the project life cycle.

REFERENCES

[1] *Project Management Body of Knowledge (PMBOK Guide)*, Program Management Institute, Newtown Square, PA, 1996, p. 6.

[2] *Building a Foundation for Tomorrow: Skills Standards for Information Technology*, Northwest Center for Emerging Technologies, Belleview, WA, 1997.

[3] *Project Management*, Univ. Washington [Online]. Available: http://www.washington.edu/computing/pm, 2003.

CHAPTER 3

Project Integration Management

Project managers that tend to focus on technical or on too many details have trouble keeping in mind and visualizing the "big picture." As stated in the previous chapter, Project managers must coordinate all of the other knowledge areas throughout a project's life cycle [1]. Table 2.4 (Chapter 2) shows that project plan development is a part of the "integration" knowledge area that takes place during the project "planning process." Project plan development entails taking the results of other planning processes and putting them into a consistent, coherent document (the project plan, also referred to as the program manager's project plan). Integration during the "executing process" entails carrying out the project plan (project plan execution), and during the "controlling process," the program manager oversees the overall change control process and ensures coordination of changes across the entire project. Interface management involves identifying and managing the points of interaction between various elements of the project.

Project managers must establish and maintain good communication and relationships across organizational interfaces.

3.1 PROJECT PLAN DEVELOPMENT

A project plan is a document used to coordinate all project planning documents. The main purpose of project plans is to "guide project execution," and to assist the project manager in leading the project team and assessing project status. Project plans are unique just as projects are unique. The main attributes of project plans are as follows:

1. They should be dynamic.
2. They should be flexible.
3. They should be updated as changes occur.
4. They should first and foremost guide project execution.

Let us reemphasize the fourth attribute. Project plans are guides that may be changed; they are not rigid regulations or laws.

The most common elements of a project management plan include an introduction or an overview of the project, a section describing how the project is organized, a section on management and technical processes used on the project, and a section describing the work to be done, a section containing the work schedule with details on the "work breakout schedule" (WBS), with budget information. Table 3.1 contains a sample outline for a project management plan (PMP). The introduction section should start with a statement of the problem as the very first sentence of the section followed by the rationale for working on the project. The section should contain full sentences and paragraphs on the project overview, background on previous or similar projects, any reference materials, what the expected outputs or deliverables, and a glossary of definitions and/or acronyms [2]. The project organization section should contain a paragraph describing the process model to be used with justification, paragraphs discussing the organizational structure, its boundaries, and interfaces. Project responsibilities should be clearly assigned and described. The section on managerial process should contain paragraphs detailing the managerial objectives, priorities, assumptions, dependencies, and constraints, and the monitoring and controlling mechanisms to be used. In addition, the section should contain detailed descriptions of the staffing plan and of the risk management process. The technical process section should contain detailed information on the method, tools, and techniques in the product development, as well as documentation on hardware and software design, drawings, operation description, and maintenance. The section should discuss project support functions. The last section contains descriptions of the work packages, dependencies, resource requirements with justifications, budget and resource allocations, and the project schedule.

3.2   PROJECT PLAN EXECUTION

Project plan execution involves managing and performing the work described in the project plan. In general, the majority of time and money is usually spent on the execution phase. The application area or the project directly affects project execution, because the products of the project are produced during execution. Project managers will use the following skills during the execution phase:

1. General management skills, which include leadership, communication, and political skills.
2. Product skills and knowledge.

In addition, the execution phase requires skills in the use of specialized tools and techniques.

Techniques to assist the project manager during project execution include the work authorization system that provides a method for ensuring that qualified people do work at the right time and in the proper sequence, and conducting "status review meetings" on a regular (weekly or monthly) schedule to exchange project information and evaluate project progress. The authors prefer weekly face-to-face meetings with monthly written status (progress) reports.

TABLE 3.1: Sample Outline for a Project Management Plan (PMP)

	INTRODUCTION	PROJECT ORGANIZATION	MANAGERIAL PROCESS	TECHNICAL PROCESS	WORK PACKAGES, SCHEDULE, AND BUDGET
Section topics	Project overview, deliverables; evolution of PMP; reference materials; definitions and acronyms	Process model; organizational structures, boundaries, and interfaces; responsibilities project responsibilities	Management objectives and priorities; assumptions, dependencies, constraints; risk management; monitoring; controlling mechanisms; staffing plan	Method, tools, and techniques; hardware documentation; software documentation; project support functions	Work packages dependencies; resource requirements; budget and allocations; schedule

To assist project managers in managing projects, there are special project management software, e.g., Microsoft Office Project. The final project process requiring the "integration" knowledge area is the "project controlling process," which requires the project manager to oversee and control product or project changes. For the medical device industry, FDA requires a paper trail of documents on any changes to a product.

3.3 PROJECT CONTROLLING PROCESS AND CHANGE CONTROL

Overall change control involves identifying, evaluating, and managing changes throughout the project life cycle. Three main objectives of change control are to

1. influence the factors that create changes to ensure they are beneficial,
2. determine that a change has occurred, and
3. manage actual changes when and as they occur.

Changes to a project or product may result from the need to take some corrective action, change requests, or from the reviews of status and progress reports. Starting with the established baseline plan or documents, status and progress reports are compared to the baselines. If it is determined that changes have occurred or should (need) to occur, then some corrective action or actions must be taken, which requires updating and documenting the update in the modified project plans. The update project plans become the current established baseline. The process is not as simple as it may appear, since government regulations mandate a formal "change control system."

3.4 CHANGE CONTROL SYSTEM

A change control system is a formal, documented process that describes when and how the official project documents and work may be changed. The change control system describes who is authorized to make changes and how to make the changes. Thus, the project manager must establish a change control board (CCB); develop a configuration management office with personnel and a configuration management plan; and establish a process for communicating changes to all stakeholders. The CCB is a formal group of people responsible for approving or rejecting changes on a project. The CCB provides guidelines for preparing or changing requests, evaluates all requests, and manages the implementation of approved changes. The board should include stakeholders from the entire organization.

Since some CCBs only meet occasionally, it may take too long for changes to occur in a timely manner. Therefore, some organizations have policies in place for time-sensitive changes.

The "48-h policy" permits project team members to make decisions, and then they have an additional 48 h to reverse the decision pending senior management approval.

3.5   CONFIGURATION MANAGEMENT

Configuration management ensures that the products and their descriptions are correct and complete. It concentrates on the management of technology by identifying and controlling the functional and physical design characteristics of products.

Configuration management specialists identify and document configuration requirements, control changes, record and report changes, and audit the products to verify conformance to requirements. Because of its importance, the authors suggest that project managers view project management as a process of constant communications and negotiations. Managers should plan for change; therefore, establish a formal change control system, including a CCB. The manager should oversee the use of good configuration management with defined procedures for making timely decisions on smaller changes. Managers should not rely solely on verbal communications, but should use written and oral performance reports to help identify and manage change. Project managers should learn to use project management and other software to help manage and communicate changes.

3.6   NEED FOR TOP MANAGEMENT COMMITMENT

Several studies cite top management commitment as one of the key factors associated with project success. A study by Pinto and Slevin in 1987 lists the key factors as

1.   top management support
2.   clear project mission
3.   good project schedule/plan
4.   good client consultation.

Whereas, the Standish Group Study (1995) lists the key factors as

1.   executive management support
2.   clear statement of requirements
3.   proper planning
4.   user involvement.

Top management can help project managers secure adequate resources, get approval for unique project needs in a timely manner, receive cooperation from people throughout the organization, and learn how to be better leaders. Project managers should meet the need for organizational standards. Senior management should encourage the use of standard forms and software for project management, the development and use of guidelines for writing project plans or providing status information, and the creation of a project management office or center of excellence. It is a well-tried and proven fact that standards and guidelines help project managers to be more effective.

REFERENCES

[1] *Project Integration, Project Management*, University of Washington [Online]. Available: http://www.washington.edu/computing/pm/plan/integration.html, 2003.

[2] *Project Plan, Management*, University of Washington [Online]. Available: http://www.washington.edu/computing/pm/plan, 2003.

CHAPTER 4

Project Scope Management

Studies in the mid-1990s cite a clear project mission with a clear statement of requirements as being important for project success, e.g., the Keller Graduate School of Management cites proper project definition and scope as the main reasons of project failure. Defining the project, product, or problem should not be limited to what is to be accomplished, but should also include what will not be accomplished; i.e., setting boundaries as to what the product is expected to do, and what the product is not being designed to do.

So what is project scope management? Project scope refers to all the work involved in creating the products of the project and the processes used to create them. It is important that project scope management include the processes involved in defining and controlling *"what is"* and/or *"what is not"* included in the project. Therefore, it is essential that the project team and stakeholders must have the same understanding of what products will be produced as a result of the project, and what processes will be used in producing them.

4.1 PROJECT SCOPE MANAGEMENT PROCESSES

Recall from Fig. 2.4 of Chapter 2 that project managers will use the project scope knowledge area throughout the project processes. Project initiation process occurs at the beginning of a project or when the project continues from a completed phase to the next phase.

The first step in initiating the projects is to look at the big picture or strategic plan of an organization. Strategic planning involves determining long-term business objectives, and it is the project managers to make sure those projects support strategic and financial business objectives.

Many organizations follow a planning process for selecting projects. The first step is to develop a strategic plan based on the organization's overall strategic plan. The second step is to perform a business area analysis, and then potential projects, project scope, benefits, and constraints are defined. The final step is to select the most viable projects and assign resources. During the planning process, the project manager develops project scope planning documents to provide the basis for future project decisions. It is during this process that the project manager develops the scope definition, subdividing the major project deliverables into smaller, more manageable components. During the executing process, the project scope verification documentation are developed and executed in formalizing acceptance of the project scope. Project scope change control documents are developed and used during the controlling process to ensure compliance in controlling changes to project scope.

4.2   SELECTING PROJECTS

There are usually more projects than the available time and resources to implement them; therefore, it is important to follow a logical process in selecting projects to work on. Methods for selecting projects to work on include focusing on broad needs, categorizing projects, financial methods, and weighted scoring models.

It is often difficult to provide strong justification for many projects, even though everyone agrees they have a high value. The "focusing on broad organizational needs" approach is based on meeting three important criteria for projects:

1.   There must be a *need* for the project.

2.   *Funds* must be available for the project.

3.   There must be a strong *will* to make the project succeed.

The "categorizing projects" approach is based on the following categories:

1.   What does the project addresses?
   a. a problem,

   b. an opportunity, or

   c. a directive for higher management.

2.   How long it will take to do the project and when it is needed?

3.   What is the overall priority of the project within the organization?

The "financial analysis of projects" approach is based on the premise that financial considerations are an important consideration in selecting projects. Three primary methods for determining the projected financial value of projects are net present value (NPV) analysis, return on investment (ROI), and payback analysis. NPV analysis is a method of calculating the expected net monetary gain or loss from a project by discounting all the expected future cash inflows and outflows to the present point in time. If financial value is a key criterion, then projects with a positive NPV should be considered: the higher the NPV, the better.

ROI is the income divided by investment, as shown in Eq. (4.1):

$$ROI = \frac{\text{total discounted benefits} - \text{total discounted costs}}{\text{discounted costs}} \qquad (4.1)$$

Most organizations have a required rate of return or minimum acceptable rate of return on investment for projects; thus, the higher the ROI, the better.

Another important financial consideration is "payback analysis." The payback period is the amount of time it will take to recoup, in the form of net cash inflows, the net dollars invested in a project. Payback occurs when the cumulative discounted benefits and costs are greater than zero. Many organizations want projects to have a fairly short payback period.

4.3   WEIGHTED SCORING MODEL

A weighted scoring model is a tool that provides a systematic process for selecting projects based on many criteria. The first step in the weighted scoring model is to identify the criteria important for the project selection process. The second step is to assign weights (percentages) to each criterion so that the total weights add up to 100%. The next step is to assemble an evaluation team, and have each member evaluate and assign scores to each criterion for each project. In the last step, the scores are multiplied by the weights and the resulting products are summed to get the total weighted scores. Projects with higher weighted scores are the best options for selection, since "the higher the weighted score, the better."

4.4   PROJECT CHARTERS

After an organization or the program manager has decided what project to work on, it is important to formalize projects with official documents. A project charter is a document that formally recognizes the existence of a project and provides direction on the project's objectives and management. It is important to have key project stakeholders and senior leadership (management) sign a project charter to acknowledge the agreement on the need and intent of the project. Either the project charter or the project management plan should contain a formal scope statement. A scope statement is a document used to develop and confirm a common understanding of the project scope, and it should include the following sections: a project justification, a brief description of the project's products, a summary of all project deliverables, and a statement of what determines project success (What are the criteria for the project's success?).

4.5   WORK BREAKDOWN STRUCTURE

After completing scope planning, the next step is to further define the work by breaking it into manageable pieces. Good scope definition helps improve the accuracy of time, cost, and resource estimates, defines a baseline for performance measurement and project control, and aids in communicating clear work responsibilities. A WBS is an outcome-oriented analysis of the work involved in a project that defines the total scope of the project.

TABLE 4.1: Example of WBS in Tabular Form

1.0 Concept
1.1 Evaluate current systems
1.2 Define requirements
    1.2.1 Define user requirements
    1.2.2 Define content requirements
    1.2.3 Define product requirements
1.3 Define specific functionality
1.4 Define risks and risk management approach
1.5 Develop project plan
1.6 Brief web development team
2.0 Product design
3.0 Product development
3.1 Product testing
4.0 Roll out
5.0 Support
6.0 Deliverables

It is a foundation document in project management, because it provides the basis for planning and managing project schedules, costs, and changes. An example of a WBS is given in Table 4.1.

4.6   APPROACHES TO DEVELOPING WORK BREAKDOWN STRUCTURES (WBSs)

There are four basic approaches to developing WBSs:

1. *The use guidelines approach*: Some organizations, like the Department of Defense (DOD), provide guidelines for preparing WBSs.

2. *The analogy approach*: It often helps to review WBSs of similar projects.

3. *The top-down approach*: Start with the largest items of the project and keep breaking them down.

4. *The bottoms-up approach*: Start with the detailed tasks and roll them up.

Most project managers will use the top-down approach and may continue to break tasks down further as the need arises.

Here are some basic principles for creating WBSs [1]:

1. A unit of work should appear at only one place in the WBS.

2. The work content of a WBS item is the sum of the WBS items below it.

3. A WBS item is the responsibility of only one individual, even though many people may be working on it.

4. The WBS must be consistent with the way in which work is actually going to be performed; it should serve the project team first and other purposes only if practical.

5. Project team members should be involved in developing the WBS to ensure consistency and buy-in.

6. Each WBS item must be documented to ensure an accurate understanding of the scope of work included and not included in that item.

7. The WBS must be a flexible tool to accommodate inevitable changes while properly maintaining control of the work content in the project according to the scope statement.

It is very difficult to create a good scope statement and WBS for a project, and it is even more difficult to verify the project scope and minimize scope changes. Many projects suffer from what is referred to as "scope creep" and poor scope verification. Scope creep occurs when additional requirements (specifications or configuration changes) are added to the project without going through an official configuration control process. Johnson [2] published a list of the top 10 factors causing project problems. The list is given in Table 4.2. The factors ranked second and third deal with inadequate scope definition and scope creep, respectively.

TABLE 4.2: Top 10 Factors Causing Project Problems

FACTOR	RANK
Lack of user input	1
Incomplete requirements and specifications	2
Changing requirements and specifications	3
Lack of executive support	4
Technology incompetence	5
Lack of resources	6
Unrealistic expectations	7
Unclear objectives	8
Unrealistic time frames	9
New technology	10

The following suggestions are offered for reducing the incomplete and changing requirements:

1. Develop and follow a requirements management process.

2. Employ techniques such as prototyping, use case modeling, and joint application design to thoroughly understand the user requirements.

3. Put all requirements in writing and create a requirements management database.

4. Use a process for reviewing requested changes from a systems perspective.

5. Provide adequate testing and emphasize completion dates.

REFERENCES

[1] D. I. Cleland, *Project Management: Strategic Design and Implementation*. New York: McGraw-Hill, 1994.

[2] J. Johnson. (1995, Jan.). CHAOS: The dollar drain of IT project failures. *Appl. Dev. Trends* [Online]. Available: www.stadishgroup.com/chaos.html.

CHAPTER 5

Personal and Project Time Management

5.1 PERSONAL TIME MANAGEMENT

Project time management is similar to personal time management. Not many young individuals (students) are very good at managing; yet it is a skill that once acquired may become a habit. I wrote a small booklet on how to manage one's time in college [1], which was used by the College of Engineering for freshmen. To compare the similarities between personal time management and project time management, let us start with the former:

> To accomplish anything, one must first have a goal; however, a goal is no more than a dream, unless you plan to accomplish the desired goal!

Setting goals should not be taken lightly, since those goals may impact one's future career and life. Therefore, be careful in determining goals and in planning how to achieve your goals. Start with an outline of what you wish to happen, and be sure to set both short-range goals and long-range goals. Then determine

1. Why are those goals necessary?
2. What are the benefits and consequences of each goal?
3. How can each goal be accomplished? (Planning)

In developing goals, be sure to make goals that are realistic, action-oriented, measurable, and include "time limits" for accomplishment of each goal. Last but not least, "prioritize the list of personal goals."

When asked, "What leads to success in achieving personal goals?" The response is, "Planning what needs to be done in order to achieve desired goals, and control, which means using time efficiently and effectively."

The next task is to create a general schedule for the week, which includes determining tasks for the week, setting priorities with "due dates," determining when to devote time to the tasks, and determining the time limits (the amount of time for each task). The first step is to fill in the working information needed for the "worksheet," then fill in all the "time blocks" scheduled for tasks (*classes* and *labs*). Next, fill in the *pre-class* and *post-class* (special study) times, and any essential times, e.g., meals, sleep, etc. Is that all? No! There is the task of managing "daily" activities. If a daily planner or organizer is used, it should be reviewed regularly, completed tasks should be crossed out, and tasks that were not completed during the week should be carried over to the next week. The final step is "implementation" of the schedule: follow the schedule and do not procrastinate.

5.2   "WORK SMARTER, NOT HARDER" [1]

In summary, to successfully manage your personal time, determine requirements for coming week. Set priorities and goals for week. Make a daily "*do-list*," and determine daily priority tasks. Review your daily "*do-lists*" in the morning before work and in the evening. Postpone unnecessary activities/tasks, and do not spread yourself too thin (learn to say, "*no!*"). When working, do only one task at a time.

5.3   PROJECT TIME MANAGEMENT

As noted in personal time management, schedules are important, and it is even more important to project managers, who indicate that delivering projects on time as one of their biggest challenges, since time schedules often have the least amount of flexibility. Unlike fictional movies, time cannot be replayed in real life. Schedule issues are the main reason for delays or conflicts on projects, especially during the execution phase of projects. Because of trying to meet some schedule, products are often introduced before all "buds" or problems with the product have been resolved. It is not surprising to read that the average time overrun on project exceeds 200%. Recently, Chip Reid, NBC Nightly News, Washington, DC (June 6, 2006) reported

> A vital $7 billion program, now approaching $11 billion, with nowhere to go, critics say, but up. A government investigation says the new polar satellite program is more than $3 billion over budget and as much as three years behind schedule. Why? The report blames "poor management oversight" by government agencies [2].

Congressional Hearing [3] on the same project was held on June 8, 2006, and was reported in the New York Times (June 9, 2006) [4]. Additionally, for those who may be interested, refer to the full audit report on the polar satellite audit by the U.S. Department of Commerce, Office of Inspector General, May 2006 [5].

5.4   PROJECT TIME MANAGEMENT PROCESSES

Project time management involves processes required to ensure timely completion of a project. The processes include

1. activity definition
2. activity sequencing
3. activity duration estimating
4. schedule development
5. schedule control.

Project schedules are developed from the basic documents that initiate a project, for example, the project charter includes start and end dates of the project with some budget information, e.g., a budget ceiling of not to exceed some target amount. The scope statement and work breakdown schedule (WBS) help define what will be done.

Activity definition involves developing a more detailed WBS and supporting explanations to understand all the work to be done; whereas, activity sequencing involves reviewing activities and determining the type of dependencies. Mandatory dependencies are inherent in the nature of the work, which are considered as hard logic; on the other hand, discretionary dependencies are defined by the project team and are considered as soft logic. External dependencies involve relationships between project and nonproject activities. In order to use critical path analysis, program managers must first determine dependencies.

5.5   PROJECT NETWORK DIAGRAMS

Project network diagrams is one technique for showing activity sequencing. A project network diagram is a schematic display of the logical relationships among project activities and/or sequencing of project activities. In the "arrow diagramming method," also called activity-on-arrow (AOA) project network diagrams, activities are represented by arrows, nodes or circles are the starting and ending points of activities. Limitation of the arrow diagramming method is that it can only show finish-to-start dependencies.

The steps in creating AOA diagrams are as follows:

1. Find all of the activities that start at the first node (node #1). Draw their finish nodes and draw arrows between node #1 and those finish nodes. Put the activity letter or name and duration estimate on the associated arrow.
2. Continuing drawing the network diagram, working from left to right. Look for bursts and merges. Bursts occur when a single node is followed by two or more activities. A merge occurs when two or more nodes precede a single node.

3. Continue drawing the project network diagram until all activities are included on the diagram that have dependencies.

4. As a rule of thumb, all arrowheads should face toward the right, and no arrows should cross on an AOA network diagram.

5.6   PRECEDENCE DIAGRAMMING METHOD (PDM)

Instead of using arrows to represent activities, the precedence diagramming method used boxes to represent activities (tasks) and arrows show relationships between activities. Many project managers will use software like Microsoft Project because of its visual approach in showing the different types of dependencies. Figure 5.1 shows how the four types of dependencies are presented in Microsoft Project.

5.7   ESTIMATION OF ACTIVITY TIMES (DURATION)

After defining activities and determining their sequence, the next step in time management is to estimate duration time for each activity. It is important to get the individuals who will be doing the actual work to help project managers create the activity estimates, and then have an expert in this area review the results.

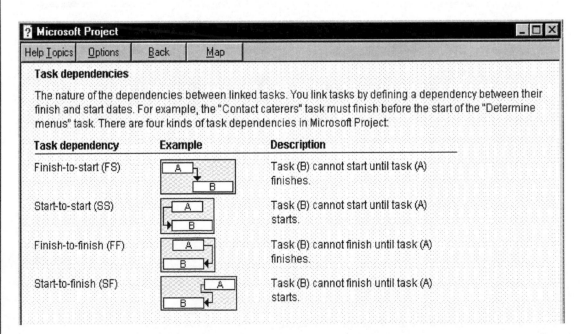

FIGURE 5.1: Activity or task dependencies. There are four types of activity dependencies: finish-to-start (FS), start-to-start (SS), finish-to-finish (FF), and start-to-finish (SF)

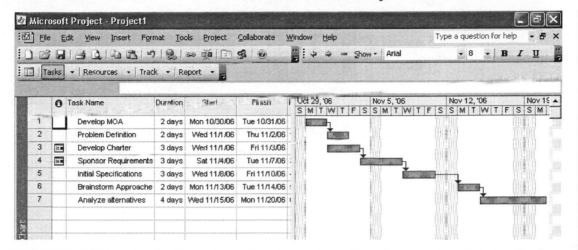

FIGURE 5.2: Example of a Gantt chart. Note that horizontal bars denote tasks and the arrows show the dependencies between the tasks

5.8   SCHEDULE DEVELOPMENT

Schedule development uses results of the other time management processes to determine the start and end dates of the project and its activities. A key challenge of project management is the creation of realistic schedules, and subsequently, to implement and stick to the schedule. The ultimate goal is to create a realistic project schedule that provides a basis for monitoring project progress for the time dimension of the project.

Important tools and techniques to assist the project manager include Gantt charts, PERT analysis, and critical path analysis. The Gantt chart was developed in 1917 by Henry Gantt as a tool for scheduling work. The Gantt chart provides a standard format for displaying project schedule information by listing project activities with corresponding start and finish dates in a calendar format. Figure 5.2 is an example of a Gantt chart. Note that horizontal bars denote tasks, and that the arrows show the dependencies between tasks. Task name and duration are shown in columns 3 and 4, where the start and finish dates are given in columns 5 and 6.

5.9   CRITICAL PATH METHOD (CPM)

The critical path method (CPM) is a project network analysis technique used to predict total project duration. A critical path for a project is the series of activities that determines the earliest time by which the project can be completed and the longest path through the network diagram, which has the least amount of slack time. If any activity on the critical path takes longer than planned, then the project schedule will slip unless corrective action is taken. There are several misconceptions about the critical path. First, the critical path is not the path that accounts for all the critical activities, since the critical path only accounts for time. Second, there may be more than one critical path in a project, if the lengths of time of more than one path are the same. Finally, the critical path is not fixed or rigid; the critical path can change as the project progresses. It is important for project managers to frequently update the project schedule information, since the critical path may change as actual start and finish dates are entered.

If the project schedule slips, then the project manager will take corrective action by applying one of the techniques for shortening a project schedule. One method in shortening durations of critical tasks is to add more resources (man hours and workers) or change the scope of the task (this may require a scope change or CCB). Another method, called "crashing," is to compress the schedule as much as possible for the least amount of incremental cost. If it is possible, the project manager may "fast track" tasks by working the tasks in parallel or overlapping them. However, if it is known that the project completion date will slip, project managers must inform stakeholders, company executives, and negotiate with the project sponsor for project time and perhaps cost overrun.

5.10   PROGRAM EVALUATION AND REVIEW TECHNIQUE (PERT)

PERT charts were developed by the Navy in 1958; however, it was not until the 1970s that the military began using project management software. PERT is a network analysis technique used to estimate project duration when there is a high degree of uncertainty about the individual activity duration estimates. PERT uses probabilistic time estimates based on using optimistic, most likely, and pessimistic estimates of activity durations. PERT uses a basic statistical weighted average formula, given by

$$\frac{\text{optimistic time} + 4 \times \text{most likely time} + \text{pessimistic time}}{6} \qquad (5.1)$$

5.11   SUMMARY

In summary, there are several hints in controlling changes to the project schedule. Project managers should perform reality checks on schedules on a regular basis and allow for contingencies. Few things work exactly as they are planned. Managers should not plan for everyone to work at 100% capacity all the time; we all need some break time. It is highly recommended that program managers hold regularly scheduled progress meetings with stakeholders, and be clear and honest in communicating schedule or project issues. In dealing with people issues, keep in mind that strong (good) leadership helps projects succeed more than all the good Gantt and/or PERT charts. Many project managers misuse project management software tools, because they do not understand important concepts of the tools and/or they have not had good training in the use of the tools or in leadership. Project managers should have good interpersonal (people) skill and should not rely solely on software tools in managing time.

REFERENCES

[1] C. Lessard, *How to Succeed in Today's TAMU*. McGraw-Hill Primis, 2001, ISBN 0-390-1-245-8.

[2] C. Reid, *NOAA Weather Satellite Program*, TV Broadcast, NBC Nightly News, Washington, DC, June 6, 2006.

[3] *NOAA Weather Satellite Program Capitol Hill Hearing Testimony*, Federal Document Clearing House, Congressional Quarterly, June 8, 2006.

[4] K. Chang, *Officials Report Progress in Weather Satellite Effort*, New York Times, Section A; Column 5; National Desk; The New York Times Company, June 9, 2006, p. 25.

[5] *Poor Management Oversight and Ineffective Incentives Leave NPOESS Program Well Over Budget and Behind Schedule*, U.S. Department of Commerce, Office of Inspector General, Audit Report no. OIG-17794-6-0001, May 2006.

CHAPTER 6

Project Cost Management

Why is project cost management so important? Technical projects in the United States have an extremely poor performance record for meeting cost goals with the average cost overrun over 189% of the original estimates and cancellation of technical projects costing the United States over $200 billion in the mid-1990s.

So, what is cost and project cost management? It is well known that to produce a product, resources in terms of personnel, materials, and finances are necessary. Cost is a resource, usually measured in monetary units like dollars, used to achieve a specific objective or given up in exchange for something. Project cost management includes the processes required to ensure that the project is completed within an approved budget (in monetary units).

6.1   PROJECT COST MANAGEMENT PROCESSES

From Table 2.4 in Chapter 2, it is noted that project cost management takes place during the planning and controlling phases of project processes. Project cost management includes the following knowledge areas during the planning process:

1. Resource planning requires the project manager to determine what resources and the amount of resources are necessary for the project.

2. Cost estimation requires the project manager to develop an estimate of the costs for the resources needed to complete a project.

3. Cost budgeting requires allocating the overall cost estimate to each individual work activity in order to establish a baseline for measuring performance.

Cost control is the knowledge area that occurs in the controlling process. Cost control requires the project manager to control changes to the project budget.

Since most chief operating officers (CEOs) and company board members may know a lot more about finance than do young engineers or project managers, it is highly recommended that one must at least learn to speak their (CEO and company officers) language or attend formal business courses in economics, accounting, and finance. Here are some simplified definitions of financial terms:

1. Profits are revenues minus expenses.
2. Project life cycle costing means determining and developing estimates for the cost of a project over its entire life.
3. Cash flow analysis means determining and developing the estimated annual costs and benefits for a project.
4. Benefits and costs can be tangible or intangible, direct or indirect.
5. Sunk cost should not be the criteria in project selection.

6.2   RESOURCE PLANNING

Since resource planning may be affected by the nature of the project and/or the organization, project managers should take into consideration the following questions:

1. How difficult will it be to accomplish specific tasks on the project?
2. What, if anything, is there in the project's scope statement that may affect resources?
3. What is the organization's past history in accomplishing similar projects or tasks?
4. Does the organization have in place the people, equipment, and materials that are capable and available for performing the work?
5. If not, can the organization acquire the necessary resources in a timely manner so as not to delay the project or tasks start times.

6.3   COST ESTIMATING

Project managers should realize that the important output of project cost management is a good cost estimate; additionally, it is important to develop a cost management plan that describes how cost variances will be managed on the project. To assist project managers in this endeavor, there are several types of tools and three techniques (rough order of magnitude (ROM), budgetary, and definitive) to help create cost estimates. Table 6.1 may help in determining when each technique may be applied in the development of a project or product.

6.4   COST ESTIMATION TECHNIQUES

Cost estimation techniques include the "top-down" or "analogous" approach that depends on using the actual cost of a previous and similar project as the basis for the new estimate. This

TABLE 6.1: Types of Cost Estimates

TYPE OF ESTIMATE	WHEN?	WHY?	ACCURACY
Rough order of magnitude (ROM)	Very early in planning phase	Rough estimate for decision selection	±25%
Budgetary	Early in the budget planning	$ into budget plans in planning phase	±10–20%
Definitive	Later in the project in execution phase	Actual cost; detail for purchasing	±5–10%

approach depends on the availability of organization archives (files) or personnel that were involved in the previous project. The "bottom-up" techniques require project managers or team to arrive at individual work item estimates without previous examples and to sum tasks estimates to obtain a total cost estimate. Additionally, there is the "parametric" technique that uses project characteristics in a mathematical model to estimate costs, e.g., the constructive cost model (COCOMO).

The developers of COCOMO contend that parametric model is the only technique that does not suffer from the limits of human decision-making. From experience, computer decision-making is only as good as the programming logic developed by humans. Flights to the moon have shown that the human decision-making could not be replaced by any automated computer program. There are computerized tools, such as spreadsheets, project management software, or other software to help project managers estimate costs.

6.5   PROBLEMS IN COST ESTIMATION

As stated previously, project cost estimates are done at various stages of the project. Developing cost estimates for a large project is a complex task requiring a significant amount of effort. A potential arises from the lack of experience of individuals attempting to develop project and/or task estimates; if this is the case, the project manager should provide cost estimation training and mentor individuals working on estimations. Another shortcoming is that most individuals have a bias toward underestimation; therefore, outside experts should be contracted to review estimates and ask important questions to make sure estimates are not biased. In many organizations, upper management wants a number for a bid, whether the final cost estimate is or is not a real estimate. The authors have experienced companies that have underestimated costs in order to obtain a contract. Project managers must negotiate with project sponsors to create "*realistic cost* estimates."

6.6   COST BUDGETING

Cost budget involves allocating the project cost estimate to individual work activities and providing a cost baseline. Most organizations have developed a standardized format to be used in developing a project budget. Deviations from the format may make it difficult to understand, evaluate, and/or compare the budget request with competing applications. The budget should contain justification stating why and how funds in each budget category are to be used. Justifications need not be elaborate, but must present a clear rationale for the use of the requested funds [1].

6.7   GUIDELINES FOR PREPARING BUDGET

Company guidelines are based on management policies found in the respective company's project manager's program guide. Guidelines should be followed when preparing the budgets and justifications. The following sections are generally included in most budgets:

A. Personnel
   (1) *Direct labor* (*salaries*): List separately by name and program title for each person to be supported by budget, list annual salary, percent of time, and the number of months to be supported.
   (2) *Benefits*: Fringe benefits are to be included in project budget requests, and the fringe benefit rate is to be presented as a part of the budget.
   (3) *Temporary help*: Fees for clerical or other staff, who are engaged on a short-term hourly basis, should be projected. List hourly rate and total hours.
   (4) *Consultants*: If the name of the consultant is known, show name and title. Indicate fees by the number of days and daily rate.

B. Supply and expense
   (1) *General expenses*: Include program and office-related expenses (e.g., photocopy expenses such as paper, copier rental, service contract, etc.).
   (2) *Communications expenses*: Include telephone and postage expenses.
   (3) *Publication costs*: Include, but do not limit to, newsletters, continuing education calendars, announcements, and educational materials you will publish or cause to be published.
   (4) *Other expenses*: Include subscriptions, books, audiovisuals, and miscellaneous expenses not covered in any of the above three categories.

C. Rental
   (1) Justification is necessary for each rental required to support the project, present monthly cost, and the number of months rented.

D. Meeting expenses

(1) Meeting funds may support planning and development of continuing professional education. Allowable costs include, but are not limited to, meeting room rental and room use charges and equipment use charge for meetings.

E. Travel

(1) Display number of trips, origin and destination, and round trip rate for airfare. Automobile usage should display total mileage and per mile rate. If per diem is requested, show the number of days and per diem rate.

F. Equipment

1. If property is to be acquired on this grant, show each item separately, indicating manufacturer or seller of the equipment, brand name, model number, and cost.

6.8   COST CONTROL

Project cost control is performed during the controlling phase of the project. As the term cost control implies, it requires the project manager to monitor "cost performance" and to ensure that only appropriate project changes are included in any necessary revision of the "cost baseline." Should there be any delays in the schedule or changes in configuration, the project manager needs to inform the project stakeholders of "authorized" changes to the project that will affect costs.

The "earned value analysis (EVA)" is an important tool used by project managers for cost control. EVA is a project performance measurement technique that integrates scope, time, and cost data [1]. Given the original planned baseline cost plus approved changes, a project manager can determine how well the project is meeting its goals: scope, time, and cost. EVA is explained in greater detail in Chapter 7.

REFERENCE

[1] Office of Management and Budgets. (2006). *Preparation and Submission of Budget Estimates*, OMB Circular no. A-11 [Online]. Available: http://www.whitehouse.gov/omb/circulars/a11/current_year/guide.pdf

CHAPTER 7

Earned Value Analysis

Earned value analysis (EVA) is an industry standard method of measuring a project's progress at any given point of time, forecasting its completion date and final cost, and analyzing variances in the schedule and budget as the project proceeds. It compares the planned amount of work with what has actually been completed, to determine if the cost, schedule, and work accomplished are progressing in accordance with the plan. As work is completed, it is considered "earned." The Office of Management and Budget prescribed in Circular A-11, Part 7, that EVA is required on construction projects:

> Agencies must use a performance-based acquisition management system, based on ANSI/EIA Standard 748, to measure achievement of the cost, schedule and performance goals [1].

EVA is a snapshot in time, which can be used as a management tool as an early warning system to detect deficient or endangered progress. It ensures a clear definition of work prior to beginning that work. It provides an objective measure of accomplishments, and an early and accurate picture of the project status. It can be as simple as tracking an elemental cost estimate breakdown as a design progresses from concept through to 100% construction documents, or it can be calculated and tracked using a series of mathematical formulae (see below). In either case, it provides a basis for course correction. It answers two key questions:

1. At the end of the project, is it likely that the cost will be less than, equal to, or greater than the original estimate?
2. Will the project likely be completed on time?

7.1   WORK BREAKDOWN STRUCTURE (WBS)

EVA works most effectively when it is compartmentalized, i.e., when the project is broken down into an organized work breakdown structure (WBS). The WBS is used as the basic building block for the planning of the project. It is a product-oriented division of project tasks that ensures the entire scope of work is captured, and allows for the integration of technical, schedule, and cost information. It breaks down all the work scope into appropriate elements for planning, budgeting, scheduling, cost accounting, work authorization, progress measuring, and management control. The two most common WBS systems are the Construction Specifications Institute (CSI) [2] format and the Uniformat II [3]. Often at the preliminary stages of design, the Uniformat II lends a better understanding of the cost centers, and at final bid level of documents, often the CSI format is used. The indirect costs of design, oversight, and management must be included in the WBS to reflect the full budget.

7.2   CALCULATING EARNED VALUE

Earned value management measures progress against a baseline. It involves calculating three key values for each activity in the work breakout schedule (WBS):

1. The planned value (PV), formerly known as the "budgeted cost of work scheduled" (BCWS) or simply called the "budget," is that portion of the approved cost estimate planned to be spent on the given activity during a given period.

2. The actual cost (AC), formerly known as the "actual cost of work performed" (ACWP) is the total of the costs incurred in accomplishing work on the activity in a given period. Actual cost must correspond to whatever activities or tasks were budgeted for the planned value and the earned value, e.g., all labor, material, equipment, and indirect costs.

3. The earned value (EV), formerly known as the "budget cost of work performed" (BCWP) is the value of the work actually completed.

These three values are combined to determine at that point of time whether or not work is being accomplished as planned. The most commonly used measures are the cost variance (CV), which is the difference between EV and AC, and is given by

$$CV = EV - AC \qquad (7.1)$$

and the schedule variance (SV), which is the difference between EV and PV or budget, is calculated as

$$SV = EV - PV \qquad (7.2)$$

These two values can be converted to efficiency indicators to reflect the cost and schedule performance of the project. The most commonly used cost-efficiency indicator is the cost performance index (CPI), which is the ratio of EV to AC, and is calculated as

$$CPI = \frac{EV}{AC} \qquad (7.3)$$

The sum of all individual EV budgets divided by the sum of all individual ACs is known as the cumulative cost performance index (CCPI) and is generally used to forecast the cost to complete a project.

The schedule performance index (SPI) is the ratio of EV to PV, and is calculated as

$$SPI = \frac{EV}{PV} \tag{7.4}$$

SPI is often used with the CPI to forecast overall project completion estimates. The general rules in interpreting EVA numbers are as follows:

1. Negative numbers for cost and schedule variance indicate problems in those respective areas.

2. A negative SV calculated at a given point of time means the project is behind schedule, while a negative CV means the project is over budget.

3. CPI and SPI less than 100% indicate problems.

7.3   EARNED VALUE MANAGEMENT SYSTEM (EVMS)

Section A-11, Part 7, of the ANSI Standard 748 [4] requires an earned value management system (EVMS) to be used to comply with the standard. A list of guidelines is provided that covers areas such as planning, scheduling and budgeting, accounting issues, management reports, and so forth; however, there are no "approved" systems identified.

The basics of any EVMS are

1. a methodical, organized, thorough, and complete WBS,

2. a baseline schedule,

3. a baseline budget, organized into control accounts,

4. measurement of the work by control account (e.g., $, units in place, man hours, etc.).

Scheduling the authorized work is no different than in any large construction project—it is a necessary activity for the success of the project. However, in an EVMS, the schedule will integrate all of the technical, cost, and schedule aspects of the work, resulting in the expected sequence of work. Interdependencies are established that result in the total work time and reveal the critical path, which is also the shortest project duration.

Within each task, it is then necessary to identify objective interim measures to allow for accurate performance assessment each month. A sufficient number of these interim measures will be defined after the detailed schedule is established to ensure the performance is measured as accurately as possible.

A time-phased budget baseline, at the control account level, must also be established and maintained. The assignment of budgets to work activities or tasks results in a plan against which actual performance can be measured. This is referred to as the performance measurement baseline (PMB), and it should be established as early as possible after a notice to proceed has been issued. The PMB includes direct hours/dollars, direct material dollars, equipment and any other direct costs, and any indirect costs for the agreed scope. The indirect costs associated with design, oversight, and management must also be included. Essentially, the PMB represents the formal plan for the project manager to accomplish all the work required in the time allotted and within the budget provided.

ANSI 748 also requires

On at least a monthly basis, generate schedule variance data that provide visibility into root causes and establish actions to achieve project completion. The first intent if this criterion is to establish the fact that analysis, to remain viable, must be accomplished on a regular, periodic basis. The second intent is to foster analyses and identification of root cause and resulting impacts at the control account level.

The monthly performance report must include

1. budget, earned value, and actual costs (reconcilable with accounting system),
2. CV,
3. SV,
4. variance at completion (VAR),
5. a variance analysis narrative (root causes, impacts at completion, and management actions).

7.4   TOOLS AND TECHNIQUES

Spreadsheets are a common tool for resource planning, cost estimating, cost budgeting, and cost control. Many organizations prefer to use more sophisticated and centralized financial applications software for cost information. Additionally, there are several software packages in the market to help the project managers prepare EVA, e.g.

1. Schedulemaker
2. Planisware OPX2
3. Risk Trak
4. Winsight
5. Primavera.

7.5   SUMMARY

Since EVA is an industry standard method of measuring a project's progress, project managers should be skilled in applying and interpreting EVA values. Project managers should not be afraid of measuring a project's progress and performance on a regular basis. Team members should be taught to use and report their respective activities performance using the EVA. Information on EVM systems is available at the Web site www.acq.osd.mil/pm.

REFERENCES

[1] Office of Management and Budgets. (2006). *Preparation and Submission of Budget Estimates*, OMB Circular no. A-11 [Online]. Available: http://www.whitehouse. gov/omb/circulars/a11/current_year/guide.pdf.

[2] Construction Specifications Institute (CSI). *WBS Format* [Online]. Available: http://www.csinet.org, 2004.

[3] Uniformat II. (1998, May). New building design management tools for project managers. *Project Manager* [Online]. Available: http://www.uniformat.com/building-design-management.html.

[4] *Earned Value Management Systems*, American National Standards Institute (ANSI)/ Electronic Industries Alliance (EIA) Standard 748-1998, May 19, 1998.

CHAPTER 8

Project Quality Management

In the past couple decades, there have appeared many articles in newspapers and on TV related to quality problems in U.S. products; not to mention the numerous jokes around the country about the poor quality of U.S. cars and computer softwares. Since the public seems to accept systems being down occasionally or needing repairs, a basic question is should we accept lower quality from newer products, with more innovation? If so, watch out for those new futuristic cars or airplanes.

Quality is defined by the International Organization for Standardization (ISO) as all the characteristics of an entity that bear on its ability to satisfy stated or implied needs. Other organizations define quality as conformity to requirements in meeting written specifications and ensuring that a product is fit to use as it was intended.

8.1 PROJECT QUALITY MANAGEMENT PROCESSES

Project quality management processes take place during the planning, execution, and control phases of project management, as shown in Chapter 2 (Fig. 2.4). Quality planning process, which takes place during the planning phase, identifies which quality standards are relevant to the project and how to satisfy them. Quality assurance is done throughout the execution phase to evaluate the overall project performance, and to ensure the project (product) satisfies the applicable quality standards while identifying ways to improve overall quality. Quality control is accomplished during the controlling phase by monitoring specific project (product) results to ensure that they comply with the relevant quality standards.

The basic requirements or objectives of quality management include

1. the requirement for customer satisfaction,
2. preference for prevention over inspection, and
3. recognizing that management has the responsibility for quality.

8.2 QUALITY PLANNING

Project managers should recognize the importance of considering quality in the very early stages of a product design and in communicating important factors that directly contribute to meeting the customer's requirements. Often during the feasibility phase, project teams may have to design experiments that can help in identifying which variables have the most influence on the overall outcome of a process. The project manager should keep in mind that many scope aspects of projects may affect quality, i.e., functionality, features, system outputs, performance, reliability, and maintainability.

8.3   QUALITY ASSURANCE

Quality assurance includes all the activities related to satisfying the relevant quality standards for a project; however, another goal of quality assurance is to provide continuous quality improvement. For example, benchmarking can be used to generate ideas for quality improvements, and quality audits can help identify lessons learned that may be used to improve performance on current or future projects.

8.4   QUALITY CONTROL

Quality control in essences requires testing or monitoring a specific product to ensure compliance with quality standards. The main outputs of quality control include making an acceptance decision on whether the product met required specifications and standard or it did not. If the product does not meet the quality standards then the product must be rejected for "rework" and/or the production process must be reviewed and perhaps require adjustments.

Some tools and techniques used in quality control include

1. Pareto analysis,
2. statistical sampling,
3. quality control charts, and
4. testing.

8.4.1   Pareto Analysis

Pareto analysis involves identifying the principle factors that account for the most quality problems in a system. Pareto analysis is also called the 80–20 rule, which means that 80% of problems are often due to 20% of the causes (factors). To help identify and prioritize problem areas in a system, Pareto diagrams or histograms are used by management personnel. Dr. Joseph Juran expanded the Pareto principle to quality issues, which is also known as the "vital few and the trivial many," thus implying that the remaining 80% of the causes should not be totally ignored [1].

TABLE 8.1: Common Certainty Factors

DESIRED CERTAINTY (%)	CERTAINTY FACTOR	SIGNIFICANCE LEVEL (α)	SAMPLE SIZE (N)
95	1.960	0.05	384
90	1.645	0.10	68
80	1.281	0.20	10

8.4.2 Quality Control Charts

Quality control charts graphically display quality data to show the results of a process over time, thus helping to determine if the process is in control or out of control, and to prevent product defects. One of the quality control charts examines the process for nonrandom problems through the application of the "Seven Run Rule," which states, "If seven data points in a row are all below the mean, above the mean, increasing, or decreasing, then the process needs to be examined for non-random problems."

8.4.3 Statistical Sampling and Standard Deviation

Statistical sampling involves choosing part (N number of samples) of a population of interest for inspection. The size of a sample (N) depends on how representative is the desired certainty. The formula for calculating the sample size is given by

$$\text{Sample size: } N = (0.25)(\text{certainty factor / acceptable error: } \alpha)^2 \qquad (8.1)$$

Table 8.1 shows several common certainty factors used in calculating the sample size. It should be noted that the significance level or acceptable error (α) is equal to one minus the desired certainty expressed in decimal format: for example, $\alpha = 0.05 = 1 - 0.95$.

Equation (8.2) is an example for calculating the sample size when the desired certainty is 95%:

$$\text{Sample size } (N) = (0.25)(1.960/0.05)^2 = 384 \qquad (8.2)$$

8.4.4 Basic Statistical Measures

Most college students are familiar with basic descriptive statistics, e.g., mean, median, mode, variance, and standard deviation. The standard deviation (σ) is a measure of how much variation exists in a distribution of data. A small standard deviation means that data are clustered closely around the middle of a distribution and that there is little variability among the data. The standard normal distribution, often referred to by students as the "bell curve," is symmetrical

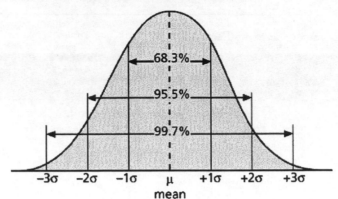

FIGURE 8.1: The standard normal distribution

about the mean or average value of a population or samples (Fig. 8.1). One should know that within $\pm 1\sigma$, 68.3% of the samples lie within the one standard deviation range, 95.5% of the samples lie within $\pm 2\sigma$, and 99.7% of the samples lie within $\pm 3\sigma$.

Quality control personnel use the terminology of "four sigma" ($\pm 4\sigma$) to indicate how good the quality of their product is. Four sigma means that only 0.0063% of the products do not meet the required standards and are therefore rejected. Table 8.2 shows the relationship between sigma and the number of defective units based on 1 billion units.

Greenberg and Hemphill [1] in a white paper contend that the main shortcoming of current quality control systems has been their inability to provide effective links to integrate with enterprise management systems.

TABLE 8.2: Relationship Between Sigma and The Number of Defective Units

SIGNIFICANCE RANGE ($\pm\sigma$)	PAPULATION WITHIN RANGE (%)	DEFECTIVE UNITS PER BILLION
± 1	68.27	317,300,000
± 2	95.45	45,400,000
± 3	99.73	2,700,000
± 4	99.9937	63,000
± 5	99.999943	57
± 6	99.9999998	2

Quality managers have long been utilizing quality control systems, including statistical process control, production part approval process, failure mode effects analysis, gage calibration and document control. But these systems traditionally have been stand-alone applications. Although these individual applications have been touted as complete quality management systems, they cannot meet all quality objectives for data collection and information sharing required for today's complex manufacturing processes [2].

8.4.5   Testing

Testing is one of the quality management functions that should be done during almost every phase of the product development life cycle; even though some managers prefer to think of testing as a stage that follows at the end of the product development process.

Test requirements and test plans are developed in the planning process (Chapter 2, Table 2.4) with the development of the quality plan well before the product is produced. The program manager should make sure that every specification is tested and that the criteria for acceptance or rejection (pass or fail) are included in the test plan.

One may think that testing is testing, but in the development of a product, there are four basic types of tests. In "unit testing," every individual component of a product is tested to ensure that the component or unit is free of defects. "Integration testing" groups components and tests the units for functionally. "Integration testing" occurs between unit and system testing or "verification testing"; then the entire system as one entity is tested in "system testing." The last set of tests is the "user acceptance testing" that has end users independently perform "validation tests" prior to accepting the delivered system.

8.5   IMPROVING PROJECT QUALITY

A goal of quality assurance is to provide continuous quality improvement. Various suggestions have been offered for improving the quality of projects which include that program managers and organizational leadership should understand the cost of quality, and promote a quality environment and frame of mind. The suggestions include focusing on organizational influences and workplace factors that may affect product quality. Additionally, project managers should use some kind of maturity model to improve product quality.

8.5.1   Maturity Models

Maturity models are frameworks for helping organization improve their processes and systems; yet there are several categories describing the type of project management maturity model. For example, the "ad hoc" maturity model refers to a project management process that may be described as disorganized, and occasionally even chaotic, since the organization has not defined systems and processes, and project success depends on individual effort. Additionally, project will have a history of chronic cost and schedule problems. With the "abbreviated" maturity model, there are some project management processes and systems in place to track cost, schedule, and scope; however, success of the project is unpredictable with cost and schedule problems becoming the norm. In the "organized" maturity model, there are standardized, documented project management processes and systems that are integrated into the rest of the organization. Project success with the organized type of maturity model is more predictable with improved cost and schedule performance. In the "managed" maturity model, management collects and uses detailed measures of the effectiveness of project management; hence, the project success is more uniform with cost and schedule performance conforming to plan. Lastly, in the "adaptive" maturity model, feedback from the project management process, and from innovative pilot ideas and/or technologies enables continuous improvement to the project processes. Project success is the norm, with continuous improvements in cost and schedule performance.

Dr. Joseph Juran, well known as a business and industrial quality "guru" (known worldwide as one of the most important twentieth century thinkers in quality management), wrote several books including the *Quality Control Handbook*, in which he outlined 10 steps for quality improvement. Juran wrote

It is most important that top management be quality-minded. In the absence of sincere manifestation of interest at the top, little will happen below [4].

8.6   COST OF QUALITY

The cost of quality is often misunderstood by those who insist that the cost of quality only includes the cost of conformance or delivering products that meet requirements and fitness for use, since those individuals have not accounted for the cost of nonconformance or taking responsibility for failures or not meeting quality expectations into the cost of quality. When examining the cost categories related to quality, five types of cost are generally cited:

1. Prevention cost, which includes the cost of planning and executing a project so that it is error-free and/or within an acceptable error range.

2. Appraisal cost, which takes into account the cost of evaluating processes and their outputs to ensure quality.

3. Internal failure cost, which is the cost incurred to correct an identified defect before the customer receives the product.

4. External failure cost, which relates to all errors not detected and corrected before delivery to the customer. Keep in mind that these costs may include recalls and liability suites.

5. Measurement and test equipment costs, which include the capital cost of equipment used to perform prevention and appraisal activities.

8.7 INTERNATIONAL ORGANIZATION FOR STANDARDIZATION

The ISO created the Quality Management System (QMS) standards in 1987. Modified in subsequent years, the ISO 9004:2000 document gives guidelines for performance improvement over and above the basic standard. The QMS is a system that outlines the policies and procedures necessary to improve and control the various processes that will ultimately lead to improved business performance and better quality control in manufacturing.

8.8 GOOD MANUFACTURING PRACTICE

According to the current Good Manufacturing Practice (GMP), medical device manufacturers should use good judgment when developing their quality system, and apply those sections of the Food and Drug Administration (FDA) Quality System (QS) Regulation that are applicable to their specific products and operations, 21 CFR 820.5 of the QS regulation. The regulation makes clear that it is the responsibility of each manufacturer to establish requirements for each type or family of devices that will result in devices that are safe and effective. Additionally, the manufacturer is responsible for establishing methods and procedures to design, produce, and distribute devices that meet the quality system requirements [5, 6].

8.9 SUMMARY

In summary, project quality management processes take place during the planning, execution, and control phases of project management. It is not only important that the project manager continuously assess project and product quality, but it is even more important that top management be quality-minded.

REFERENCES

[1] J. M. Juran. *Pareto Principle* [Online]. Available: http://en.wikipedia.org/wiki/Dr._Joseph_Moses_Juran#Pareto_principle, 2006.

[2] N. Greenberg and L. Hemphill. *A New Approach: Enterprisewide Quality Management Systems*, White Paper, DATANET Quality Systems [Online]. Available: http://www.winspc.com/whitepapers.htm, 2005.

[3] DATANET Quality Systems. *Quality Digest Magazine* [Online]. Available: http://www.winspc.com/quality-management-systems.htm, 2005.

[4] J. M. Juran. (1945). *Quality Control Handbook* [Online]. Available: http://en.wikipedia.org/wiki/Dr._Joseph_Moses_Juran.

[5]   Quality Management Wikipedia. *The Free Encyclopedia* [Online]. Available: http://en.wikipedia.org/wiki/Quality_management, 2006.

[6]   Quality Management Wikipedia. *The Free Encyclopedia* [Online]. Available: http://en.wikipedia.org/wiki/Quality_Management_System, 2006.

CHAPTER 9

Project Procurement Management

So far, Chapters 4 through 8 have covered the four core knowledge areas: scope, time, cost, and quality management, respectively; Chapters 9 through 12 will cover the four facilitating knowledge areas: procurement management, human resources, communication, and risk. This chapter focuses on project procurement management. The term "procurement" generally means acquiring goods and/or services from an outside source; however, other terms such as "purchasing" or "outsourcing" are often used interchangeably to mean procurement. Often an organization will outsource or purchase components or subunits for another source (company) for various reasons, such as possible reduction in fixed and recurrent costs, to increase accountability, to provide the flexibility for the organization to focus on its core business, or to gain access to technical skills, expertise, and/or technologies that the organization does not possess [1].

9.1   PROJECT PROCUREMENT MANAGEMENT PROCESSES

The project procurement management processes, as shown in Table 9.1, include the following:

1. Procurement planning that takes place during the "project planning process" to determine what part or systems to procure and when to make the purchases.

2. Solicitation planning that takes place during the "project planning process" to document product requirements (materials, parts, components, etc.) and to identify potential sources (vendors) for procurement of required parts, etc.

3. Solicitation, which usually occurs during the "project executing process" to obtain quotations, bids, offers, or proposals as appropriate from the vendors.

4. Source selection that takes places after solicitation during the "project executing process" to select the best offer for the best product from potential vendors.

5. Involving "contract administration" to manage the contractual relationship with the vendor. Keep in mind that engineers generally are not trained in contractual and legal matters. "Leave contractual matters to qualified contract administrators!"

TABLE 9.1: Project Procurement Processes (Excerpt from Chapter 2, Table 2.4)

PROJECT PROCESSES				
KNOWLEDGE AREA	INITIAL-IZING	PLANNING	EXECUT-ING	CONTROL-LING	CLOSING
Procurement		Procurement planning	Solicitation, source selection		Contract closeout
		Solicitation planning	Contract administration		

6. The final project procurement process, contract closeout, which usually occurs during the "project closing process" as a formal process at the completion and settlement of the contract.

9.2   PROCUREMENT PLANNING

Procurement planning involves identifying which project needs can be best met by using products or services outside the organization. Thus, the project manager and his team must evaluate various alternatives and decide whether to make or buy. "Make-or-buy analysis" is a process used to determine whether a particular product or service should be made inside the organization or purchased from some source outside of the organization. Often, the make-or-buy analysis involves doing some financial analysis. Additionally, they must decide on "what," "how much," and "when" to procure or purchase. The purchasing decision must include "when" to procure so that the required materials or parts are on hand so as not to cause delays in the project schedule. From experience, I was told by the manufacturer of a part that they had a 52-week backlog; meaning that the project would be delayed about a year if I insisted on that part. The only available option at that time was to redesign the product with an equivalent component. If members of the project team are not familiar with the procurement process, the project manager should seek experts within the organization and/or from consultants outside of the organization that can provide valuable and experienced inputs in procurement decisions.

9.2.1   Types of Contracts

There are three basic types of contracts: fixed price, cost reimbursable, and unit price contract [2]:

1. The "fixed price contract," also referred to as the "lump sum contract," infers that the contract has a fixed total price for a well-defined product or service. A "fixed price

TABLE 9.2: Contract Types and Associated Risk		
BUYER RISK	**TYPE OF CONTRACT**	**VENDOR RISK**
Low	Fixed price	High
Medium low	Fixed price incentive (FPI)	Medium high
Medium	Cost plus insentive fee (CPIF)	Medium
Medium high	Cost plus fixed fee (CPFF)	Medium low
High	Cost plus percentage of costs (CPPC)	Low

incentive" contract is similar to the fixed price contract with the exception that the buyer will pay some incentive or bonus if the seller performs better than that written into the contract (especially, time; i.e., if the end product is produced and delivered earlier than schedule).

2. The "cost reimbursable contract" involves payment to the seller for direct and indirect costs. There are three different types of contracts within the cost reimbursable framework: cost plus incentive fee (CPIF), cost plus fixed fee (CPFF), and cost plus percentage of costs (CPPC):

   a. With a CPIF contract, the buyer pays the seller for allowable performance costs plus a predetermined fee and an incentive bonus.

   b. With the CPFF contract, the buyer pays the seller for allowable performance costs plus a fixed fee payment that is usually based on a percentage of estimated costs.

   c. With the CPPC contract, the buyer pays the seller for allowable performance costs plus a predetermined percentage based on total costs.

3. The unit price contract requires the buyer to pay the seller a predetermined amount per unit of service.

All contracts involve some risks for both the vendor (seller) and the buyer. Table 9.2 summarizes the risks associated with each type of contract. Note that the lowest risk associated with the type of contract to the buyer is a fixed priced contract; however, the vendor may not negotiate this type of contract, since the fixed price contract presents the highest risk to the vendor.

9.3 SOLICITATION PLANNING

In solicitation planning, several documents must be prepared by the procurement team. The first document, called the "request for proposals (RFP)," is used to solicit proposals from prospective sellers where there are several ways to meet the sellers' needs. On the other hand, "requests for quotes (RFQ)" are used to solicit quotes for well-defined procurements invitations for bid or negotiation in which initial contractor responses are also part of solicitation planning.

An RFP usually includes sections on the purpose of the RFP, the organization's background, requirements, environments, statement of work (SOW) with schedule, and required deliverables with schedule. Almost all mutually binding agreements or contracts include a SOW. Additionally, the contracting office will add boilerplate information required by law.

9.3.1   Statement of Work (SOW)

The SOW is usually developed by the engineering members of the procurement team. It is a description of the work required for the procurement; hence, a good SOW gives bidders a better understanding of the buyer's expectations. General format for a SOW include sections on the scope and location of work, period of work to be performed (usually, end dates) with scheduled deliverables.

The scope of work should describe in as much detail as possible the exact nature of work to be accomplished by the contractor. Not only should the hardware and software be specified, but also the required tolerances and/or industry or ISO standards to be met. If the work must be performed in a specific location, such as a designated standard clean room where employees must perform the work on hardware or a secure location for work, then the environment and location of work must be described in the SOW. Contacts often specify within the SOW when the work is expected to start and end, working hours, number of hours that can be billed per week, where the work must be performed, and related schedule information. Deliverables schedule list specific deliverables, describe what is to be delivered in detail, and when the deliverables are due.

9.4   SOLICITATION

Solicitation is a function that occurs in the executing process, and involves obtaining proposals or bids from prospective sellers in response to an RFP or RFQ. Organizations can advertise to procure goods and services in several ways: advertising to anyone that may be interested via some publication or announcement; for example, government agencies place their solicitation in the commerce daily bulletin or on their respective Web sites. Formal evaluation procedures for selecting vendors should be developed and documented before solicitation. Depending on the price and total spending, some organization may approach several potential vendors for quotations or the purchasing agency may only approach preferred vendors, often referred to as the "buyers short list". It is not unusual in large complex procurement to host a bidders' conference to help clarify the buyer's expectations, thus reducing the number of nonresponsive responses.

Responses to RFPs or RFQs always have a cutoff date, after which time any responses arriving are considered "nonresponsive" and are not evaluated.

9.5   SOURCE SELECTION

After receiving bidders' responses, procurement must assemble a source selection committee. Source selection involves evaluation of bidders' proposals and selection of the best proposal. From this point on, the purchasing department with their contract specialist (contract administrator) negotiate the contract with terms and conditions, and award the contract.

9.6   CONTRACT ADMINISTRATION

Contract administration ensures that the seller's performance meets the contractual requirements. All contracts are legal relationships between purchasing organizations and selling or service organizations, so it is important that legal and contracting professionals be involved in writing and administering contracts. Project managers and members of the design team should not make comments that may be misinterpreted as redirection of the contract terms or SOW. On the other hand, those project managers who ignore contractual issues may find that those issues result in serious problems. Project managers should be aware that changes to any part of the project including contract changes need to be reviewed, approved, and documented in the same way that the original part of the plan was approved. Evaluation of any changes to the project should include an impact analysis, "How will the change affect the scope, time, cost, and quality of the goods or services being provided?" Along with documenting in writing any changes, project team members must document all important meetings and telephone phone calls that deal with project matters.

9.7   CONTRACT CLOSEOUT

The final process in project procurement management is the formal closing of contracts. Contract closeout includes verification to determine if all work was completed correctly and satisfactorily. If there were any discrepancies or deficiencies, those deficiencies have to be dealt with either correction or wavering of the deficiencies. Contract administration is required to update records of administrative activities to reflect final contract results and to archive contract information for future use. Usually, procurement audits are performed to identify "lessons learned" in the procurement process.

REFERENCES

[1]   Procurement, Wikipedia. *The Free Encyclopedia* [Online]. Available: http://en. wikipedia.org/wiki/Procurement, 2005.

[2]   J. Bronzino, *Assessment and Acquisition, Management of Medical Technology.* London: Butterworth–Heinemann, 1992, ch. 4, pp. 111–152.

CHAPTER 10

Project Human Resource Management

So, what is project human resource management and why is this area important to project managers? Project human resource management may be defined as the processes of making the most effective use of the people involved with a project. Basically, any human resource management implies having "the right people to the right place at the right time," which requires organizational planning as to the type of personnel (engineers, support staff, etc.) that must be recruited or reassigned from within the organization and/or hired if the required talents or skills are not within the organization to work on the project. Subsequently, those resources (humans) must be developed or molded into an effective project team [1].

Having served on the Institute of Electrical and Electronic Engineers (IEEE) Workforce Committee and the American Association of Engineers Workforce Commission for a decade, the author can assert, "people determine the success and failure of projects and organizations."

The IEEE and the Bureau of Labor and Statistics have cited for the past decade shortages of trained engineers to fill between one-fourth and one-third of a million jobs openings in engineering, which makes human resource management even more challenging for projects. Many CEOs listed the lack of highly skilled, trained workers as the primary barrier to growth; hence, they lobby the U.S. Congress to increase the annual H1B immigration quota to over one-third of a million foreign immigrant workers.

Congress, universities, and many technical societies have wrestled with the problem of "How to increase the U.S. engineering labor pool." The consensus of the various organizations points to the undesirable stereotyping of engineers as "nerds" as a factor in keeping the U.S. students away from the engineering career field. Further stereotyping the noted problems of engineering is hard work requiring higher level math than simple addition and subtraction, long work hours (days), and constantly staying abreast of changes in the field. Stereotyping engineering disciplines as male-dominated tends to keep women from entering the engineering career field. Problems of shortages and reduced numbers of young engineers entering into the human resource pool means that there is a need for better human resource management within organizations.

10.1   MANAGING PEOPLE

Project managers should have not only some formal training in managing people, but also field experience in managing people at work. Important knowledge areas related to project management include

1. motivation

2. influence and power

3. effectiveness.

Maslow developed a theory that people's behaviors are guided by a sequence of needs, and he argued that humans possess unique qualities that enable them to make independent choices, thus giving them control of their destiny [2]. Maslow's *Hierarchy of Needs* starts with the need to satisfy physiological needs as the lowest motivator. One may think of these needs as the survival mode, where satisfying hunger and thirst to survive is the paramount need. From the lowest motivator to the highest, the needs to satisfy are physiological, safety, social, esteem, and self-actualization. Maslow contends that growth motives (being motives) are relatively independent of the environment and are unique to the individual [3]. He states that

> "The esteem needs usually act as motivators only if the three lower types have been satisfied to some degree."

Maslow cautions that true self-esteem is based on real competence and significant achievement, rather than on external fame. The highest form of motivation is the need for self-actualization.

Herzberg [4] distinguishes between "motivational factors" and "hygiene factors." Motivational factors include achievement, recognition, the work itself, responsibility, advancement, and growth, which produce job satisfaction. Examples of motivation factors include higher salaries, more supervision responsibilities, and a more attractive work environment. On the other hand, hygiene factors cause dissatisfaction if not present, but do not motivate workers to do more.

Thamhain and Wilemon [5] list nine ways in which project managers have to influence projects: these ways or methods include the following:

1. *Authority*: The project manager's legitimate hierarchical right to issue orders.

2. *Assignment*: The project manager's perceived ability to influence a worker's later work assignments.

3. *Budget*: The project manager's perceived ability to authorize others to use discretionary funds.

4. *Promotion*: The project manager's ability to improve a worker's position.

5. *Money*: The project manager's ability to increase a worker's pay and benefits.

6. *Penalty*: The project manager's perceived ability to dispense or cause punishment.

7. *Work challenge*: The project manager's ability to assign work that capitalizes on a worker's enjoyment of doing a particular task

8. *Expertise*: The project manager's perceived special knowledge that others deem important.

9. *Friendship*: The project manager's ability to establish friendly personal relationships between the project manager and others.

One should keep in mind that projects are more likely to succeed when project managers influence with expertise and work challenges; whereas projects are more likely to fail when project managers rely too heavily on authority, money, and penalty [6].

Power is defined as the potential ability to influence behavior to get people to do things they would not otherwise do. There are several types of power including legitimate, expert, reward, coercive (with treat of penalty or punishment), and referent (meaning to refer to a decision/problem to someone).

10.2   IMPROVING EFFECTIVENESS: COVEY'S SEVEN HABITS

Covey first published *The Seven Habits of Highly Effective People* in 1989 and the 15th anniversary edition in 2004. The book lists seven principles that, if established as habits, Covey contends, are supposed to help a person achieve true interdependent "effectiveness" [7].

10.2.1 The Seven Habits

Covey presents his teachings in a series of habits—a progression from dependence, to independence, to interdependence. The seven habits are as follows:

Habit 1: Be Proactive: Principles of Personal Vision
Habit 2: Begin with the End in Mind: Principles of Personal Leadership
Habit 3: Put First Things First: Principles of Personal Management
Habit 4: Think Win/Win: Principles of Interpersonal Leadership
Habit 5: Seek First to Understand, Then to be Understood: Principles of Empathetic Communication
Habit 6: Synergize: Principles of Creative Communication
Habit 7: Sharpen the Saw: Principles of Balanced Self-Renewal

Expansion of Covey's habits are quoted from the Wikipedia Web site (http://en.wikipedia.org/wiki/Stephen_Covey) [8].

1. Be Proactive. Here, Covey emphasizes the original sense of the term "proactive" as coined by Victor Frank. Being "proactive" means taking responsibility for everything in life, rather than blaming other people and circumstances for obstacles or problems. Initiative and taking action will then follow (the authors of this book also extend the meaning to include thinking of potential problem areas before they occur and planning alternative solutions prior to the occurrence of the bad event.) [8].

2. Begin with the End in Mind, which deals with setting long-term goals based on "true-north principles." Covey recommends formulating a "personal mission statement" to document one's perception of one's own purpose in life. He sees visualization as an important tool to develop this. He also deals with organizational mission statements, which he claims to be more effective if developed and supported by all members of an organization, rather than being prescribed [8].

3. Put "First Things First". Covey describes a framework for prioritizing work that is aimed at long-term goals, at the expense of tasks that appear to be urgent, but are in fact less important. Delegation is presented as an important part of time management. Successful delegation, according to Covey, focuses on results and benchmarks that are to be agreed in advance, rather than on prescribing detailed work plans [8].

4. Think Win/Win describes an attitude whereby mutually beneficial solutions are sought, that satisfy the needs of oneself as well as others, or, in the case of a conflict, both parties involved [8].

5. Seek First to Understand, then to be Understood. Covey warns that giving out advice before having empathetically understood a person and their situation will likely result in that advice being rejected. Thoroughly listening to another person's concerns instead of reading out your own autobiography is purported to increase the chance of establishing a working communication [8].
   Good project managers are empathic listeners; they listen with the intent to understand. Before one can communicate with others, a rapport with the other individual should be established; e.g., a social gathering or a meal so as to get to know the other person on a nonbusiness, more personal manner. Some times mirroring is a technique to help establish rapport. Project managers need to develop empathic listening and other people's skills to improve relationships with users and other stakeholders.

6. Synergize describes a way of working in teams. Apply effective problem solving. Apply collaborative decision making. Value differences. Build on divergent strengths.

Leverage creative collaboration. Embrace and leverage innovation. It is put forth that, when this is pursued as a habit, the result of the teamwork will exceed the sum of what each of the members could have achieved on their own. The whole is greater than the sum of its parts [8].

7. Sharpen the saw focuses on balanced self-renewal. Regaining what Covey calls "productive capacity" by engaging in carefully selected recreational activities [8].

10.2.2 Personality and Behavioral Tools

There are several personality and behavioral tools that help human resource managers, and could help project managers in developing an effective working team. The Meyers–Briggs type indicator (MBTI) is a popular tool for determining personality preferences and helping teammates understand each other. The MBTI has four dimensions in which individuals are classified as either an extrovert (E) or introvert (I), as sensation (S) or intuition (N), as thinking (T) or feeling (F), and as using judgment (J) or perception (P). Most professionals seem to fall within the category of intuition and thinking (NTs) or as being rationals.

Based on the work by Charles Marston (1928), the Texas A&M University Employee Assistance Program Office developed a Behavior Profile tool, which they called "DISC" [9]. The acronym DISC stands for

1. dominance, which pertains to how individuals respond to problems or challenges,
2. influence, which is defined as how individuals influence contacts with others by changing their point of view to your point of the individual,
3. steadiness, which deals with consistency on how individuals respond to the pace of the environment,
4. compliance, which addresses the issue of constraints or how individuals respond to rules and procedures set by others.

The DISC sectors denoting behavioral quadrants are shown in Fig. 10.1 [9]. The DISC behavior profiles with comments are shown in Fig. 10.2.

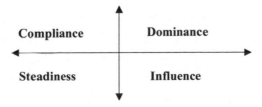

FIGURE 10.1: DISC. Quadrants show the dominant behavior profile

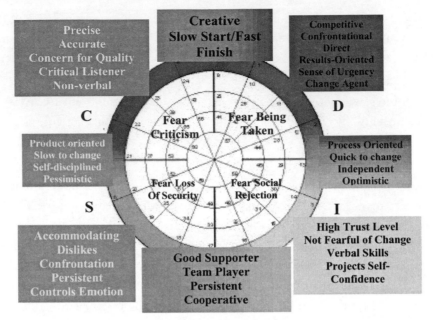

FIGURE 10.2: Behavior Profile tool, called the "DISC" [9]

10.3 SUMMARY

Repeating the advice given in Chapter 1, project managers and their teams should focus their primary attention and efforts on meeting project objectives and producing positive results. It is recommended that instead of blaming team members, the focus should be turned to fixing the problem. Project managers should establish regular, effective meetings with agenda and openness. Part of the program managers tasks include nurturing team members, encouraging them to help each other, and to acknowledge in public individual and group accomplishments.

Project managers should remember to have a team-based reward and recognition systems, since such consideration of others can promote teamwork. Also give some thought on rewarding teams for achieving specific goals. Additionally, project managers should allow time for team members to mentor and help each other meet project goals and develop interpersonal skills.

REFERENCES

[1] *Resource Assessment, Program Management*, University of Washington [Online]. Available: http://www.washington.edu/computing/pm/plan/resource.html, 2005.

[2] A. H. Maslow (2003). Self-actualization theory (II). *An Introduction to Theories of Personality (6th ed.)*, R. B. Ewen, Ed. Houston, TX: Questia Media America, Inc. [Online]. ch. 10. Available: www.questia.com.

[3] A. H. Maslow, *Motivation and Personality*, 2nd ed. New York: Harper and Row, 1970.

[4] F. Herzberg, B. Mausner, and B. B. Snyderman, *The Motivation to Work*, 2nd ed. New York: Wiley, 1959.

[5] H. Thamhain and D. L. Wilemon, "One more time: How do you motivate employees?," *Harvard Business Rev.*, pp. 51–62, 1968.

[6] B. Nath. *CSS*, University of Melbourne [Online]. Available: http://www.cs.mu.oz au/443/slides/443Lec12.pdf.

[7] S. R. Covey. (1989 and 2004). *The Seven Habits of Highly Effective People* (Paperback Publication, ISBN 0-671-70863-5) [Online]. Available: http://en.wikipedia.org/wiki/The_Seven_Habits_of_Highly_Effective_People.

[8] *The 7 Habits of Highly Effective People* [Online]. Available: http://en.wikipedia.org/wiki/Stephen_Covey, 2005.

[9] *DISC Behavior Profile*, Texas A&M University Employee Assistance Program Office, Texas A&M University, College Station, TX, 1997.

CHAPTER 11

Project Communications Management

It is interesting to note that our society and culture do not portray engineers as good communicators. Both the *New York Times* and the *Washington DC* have printed front page articles on the poor communications skills of engineers. Without a doubt, history has shown that engineers must be able to communicate effectively to succeed in their positions, since strong verbal skills are a key factor in career advancement for engineering professionals. The IEEE USA Professional Activities has published newsletter with the fact that being technically qualified is insufficient to maintain employment, the engineer and specially engineering managers must possess interpersonal and communications skills. Project managers are taught that one of the greatest threats to many projects is a failure to communicate.

11.1 PROJECT COMMUNICATIONS MANAGEMENT PROCESSES

As shown in Table 11.1, the project communications management processes occur throughout the project development phases, from planning through closing. During the project planning phase, the planning teams must determine the information and communications needs of the stakeholders, then develop and document a communications plan. During the project executing process, information on the project is collected and distributed; making sure that the necessary or needed information is made available to stakeholders in a timely manner. In the controlling process of the project, performance information are collected, analyzed, and disseminated in performance reports. Administrative closure, which occurs during the closing process, consists of generating, gathering, and disseminating information to formalize phase or project completion.

11.2 COMMUNICATIONS PLANNING

The communications management plan is a document that guides project communications. As an aid in communications planning, the project manager should create a stakeholder analysis

TABLE 11.1: Project Communications Management Processes

KNOWLEDGE AREA	INITIALIZING	PLANNING	EXECUTING	CONTROLLING	CLOSING
Communications		Communications planning	Information distribution	Performance reporting	Administrative closure

TABLE 11.2: Example of Stakeholder Analysis for Project Communications

STAKEHOLDER	DOCUMENT	FORMAT	CONTACT PERSON	DUE
Customer management	Monthly status report	E-mail and hard copy	List name and phone number	First of month

for project communications. It should go unsaid that every project should include some type of communications management plan. Communications management plans usually contain

1. a description of a collection and filing structure for gathering and storing various types of information,
2. a distribution structure describing what information goes to whom, when, and how,
3. a format for communicating key project information; formats facilitate a better understanding of the communications,
4. a project schedule for producing the information,
5. access methods for obtaining desired information,
6. a method for updating the communications management plans as the project progresses and develops,
7. and last but not least, a stakeholder communications analysis [1].

Table 11.2 is a short example of stakeholder communications analysis format. The table rows would describe the required documents, format, and due date for each stakeholder.

11.3   INFORMATION DISTRIBUTION
Getting the right information to the right people at the right time and in a useful format is just as important as developing the information in the first place. Project managers should consider

the importance of using current electronic communication technologies to enhance informa-tion distribution, as well as considering different formal and informal methods for distributing information. Project teams should know and understand the organization's communications infrastructure, which are a set of tools, techniques, and principles that provide a foundation for the effective transfer of information. The tools may include e-mail, project management soft-ware, groupware, fax machines, telephones, teleconferencing systems, document management systems, and word processors. The techniques may include reporting guidelines and templates, meeting ground rules and procedures, decision-making processes, problem-solving approaches, and conflicting resolution and negotiation techniques. Additionally, the principles foundation for the effective transfer of information should include an agreed upon work ethic and the use of free, honest, and open dialog.

11.4   SPAN OF CONTROL

Most managers know how difficult it is to control a large group of individuals. Training in military leadership and experiences in managing projects have taught the authors that for good management of people, the span of control lie between the minimum of 5 and the maximum of 12 individuals. Managing more than 12 requires the manager to delegate the responsibilities and authority (full power) to submanagers, so they may be in full control of their responsible areas.

11.5   PERFORMANCE REPORTING

Performance reporting keeps higher level management and stakeholders informed about how resources are being used to achieve project objectives. "Status" reports describe where the project stands at a specific point in time, whereas "progress" reports describe what the project team has accomplished during a certain period of time. Performance reporting should include earned value analysis (EVA) and project forecasting to predict what the project status and/or progress will be in the future based on past information and trends analysis. Project managers should hold status review meetings often (weekly or monthly) and include performance reporting (EVA) in all reports. The final process in communication management occurs during the project closure phase, and is referred to as administrative closure, which produces the following documentation:

1. project archives
2. formal acceptance
3. lessons learned.

In summary, a template for a team's "weekly progress report" is given.

11.5.1 Template for Weekly Progress Report

1. *Accomplishments for past week (include appropriate dates):*

    a. Detailed description of accomplishments. Who did the work? Relate accomplishments to project's Gantt chart by task number (refer to task number).

    b. If any issues were resolved from a previous report, list them as accomplishments.

    c. Write in full English sentences/paragraphs, not just bullets.

2. *Plans for coming week (include appropriate dates):*

    a. Detailed description of planned work task items to be accomplished in the next week.

    b. Again, who is going to do the work and relate to the project's Gantt chart by task?

    c. Describe any other unplanned items to accomplish, which are not on the Gantt tasks.

    d. Write in full English sentences/paragraphs, not just bullets.

3. *Issues:*

    a. Discuss issues (problems encountered) that surfaced or are still important.

    b. If problem encountered, discuss impact on schedule (time line), funding/cost, and proposed possible remedy (how does the team plan to get back on schedule?).

    c. Managers do *not* like surprises, so be sure to discuss issues in detail.

4. *Project changes (date and description):*

    a. List any approved or requested changes to the project.

    b. Include the date of the change and a detailed description with the reason (why?).

    c. Include proposed changes to a new Gant chart and schedule.

5. *Attachments:*

    a. Previous Gant Chart (before Update); Previous Budget (before Update),

    b. Updated Gant Chart (Time Line Update); Updated Budget,

    c. Excel Chart of Time Sheet (Hours worked).

REFERENCE

[1] *Communication Plan, Project Management,* University of Washington [Online]. Available: http://www.washington.edu/computing/pm/plan/communication.html, 2002.

CHAPTER 12

Project Risk Management

Risk is defined in the dictionary [1] as, "The possibility of loss or injury" and often expressed in terms of severity and probability. However, Project Risk Management is defined as the art and science of identifying, assigning, and responding to risk throughout the life of a project and in the best interests of meeting project objectives. Project risk involves understanding potential problems that might occur on the project and how those problems could impede project success. Risk management is often overlooked on projects by executive managers, but it can help project managers improve project success by helping select good projects, determining project scope, and developing realistic estimates. Thus, risk management may be considered as a form of an investment or insurance.

Government organizations define the terms in a slightly different manner. For example:

1. Hazard is defined as a condition with the potential to cause personal injury or death, property damage, or degradation of performance.
2. Risk is an expression of possible loss in terms of severity and probability.
3. Severity is the worst credible consequence that can occur as a result of a hazard.
4. Probability is the likelihood that a hazard will result in a mishap or loss.
5. Risk Assessment is the process of detecting hazards and assessing associated risks.
6. Control is a method for reducing risk for an identified hazard by lowering the probability of occurrence, decreasing potential severity, or both.

Risk management in the Department of Defense is referred to as Operations Risk management and is defined as the process of dealing with risks associated with daily operations, which includes risk assessment, risk decision-making, and implementation of effective risk controls [2].

12.1 PROJECT RISK MANAGEMENT
Project Risk Management is the process of identifying, assigning, and responding to risks associated with a project, rather than operations. One may question why organizations would want

to venture into a risky project, and the response would be because the opportunities outweighed the risk. So, "What is Project Risk Management?" The goal of project risk management is to minimize potential risks while maximizing potential opportunities. Major risk management processes include:

1. Risk identification, which is the process of determining risks that are likely to affect a project.

2. Risk quantification, which requires evaluating risks to assess the range of possible project outcomes.

3. Risk response development, which includes taking steps to enhance opportunities and developing responses to each threat.

4. Risk response control is the process of responding to risks over the course of the project.

Risk identification is the process of understanding what potential unsatisfactory outcomes are associated with a particular project.

12.2   TYPES OF PROJECT RISKS

There are three types of risks associated with commercial ventures in developing a new product; market risk, financial risk, and technology risk. Market risk is associated with determining if the new product will be useful to the organization or marketable to others and if users will accept and use the product or service? Financial risk is associated with determining if the organization can afford to undertake the project and determining if this project is the best way to use the company's financial resources. Technology risk is associated with determining if the project is technically feasible and if the technology could become obsolete before a useful product can be produced? Table 12.1 is a short list of the potential risk conditions that are associated with each of the knowledge areas.

12.3   RISK QUANTIFICATION

Risk quantification or risk analysis is the process of evaluating risks to assess the range of possible project outcomes. The first step in the risk analysis process is to determine the risk probability of occurrence and its impact (consequences) to the project if the risk does occur. Risk quantification techniques include expected monetary value analysis (EVA), calculation of risk factors, PERT estimations, simulations, and expert judgment. Risk simulation uses a representation or model of a system to analyze the expected behavior or performance of the system. The Monte Carlo analysis simulates a model's outcome many times in order to provide

TABLE 12.1: Potential Risk Conditions Associated with Knowledge Areas

KNOWLEDGE AREAS	RISK CONDITIONS
Integration	Inadequate planning; poor resource allocation; poor integration management; lack of postproject review
Scope	Poor definition of scope or work packages; incomplete definition of quality managements; inadequate scope control
Time	Errors in estimating time or resource availability; poor allocation and management of float; early release of competitive products
Cost	Estimating errors; inadequate productivity, cost, change, or contingency control; poor maintenance, security, purchasing, etc.
Quality	Poor attitude toward quality; substandard design/materials/workmanship; inadequate quality assurance program
Human resources	Poor conflict management; poor project organization and definition of responsibilities; absence of leadership
Communications	Carelessness in planning or communicating; lack of consultation with key stakeholders
Risk	Ignoring risk; unclear assignment of risk; poor insurance management
Procurement	Unenforceable conditions or contract clauses; adversarial relation

a statistical distribution of the calculated results. Some organizations rely on the experience of experts to help identify potential project risks. If the organization uses a number of experts, then the Delphi method is used to derive a consensus among a panel of experts in deriving predictions about future developments [3].

12.4   RISK RESPONSES

When faced with a hazard or potential risk, individuals and organizations will respond in one of three ways.

1. Risk avoidance response tries to eliminate a specific threat or risk, by eliminating its causes. The approach may take a lot of time, since to find the cause will require an in-depth look at the process with a detailed analysis. The cause should subsequently be fixed. Dorfman [4] defines avoidance as not performing an activity that could carry risk. Avoidance may seem the answer to all risks, but avoiding risks also means losing out on the potential gain that accepting (retaining) the risk may have allowed. Not entering a business to avoid the risk of loss also avoids the possibility of earning profits.

2. Risk acceptance response means accepting the consequences should a risk occur.

3. Risk mitigation response attempts to reduce the impact of a risk event by reducing the probability of its occurrence.

12.5   CAUSES OF RISK

It would seem to some individuals that risk avoidance should be the only response to a hazard or a risk; so, let us examine the causes. Change is considered the "Mother" of Risk, meaning that changes in any project will increase the risk and in some cases significantly. To change, one can add development of new technology, complexity of the system, resource constraints, organizational constraints, speed or tempo of work, environmental influences, work requiring high energy levels, and human nature [2]. The changes are not given in any order, but keep in mind that two or more causes may combine or result in one occurrence of a hazard or a risk. Table 12.2 shows strategies used by organizations to mitigate the various types of risks. To read the table, start with a type of risk, then read the optional mitigation strategies in that vertical column.

12.6   RISK MANAGEMENT PLANS

Project managers should develop three different risk management plans. The "Risk Management Plan" documents the procedures for managing risk throughout the project, but the project manager should also develop "Contingency Plans" that predefine actions the project team will take if an identified risk event occurs. Additionally, the project manager should have "Contingency Reserves," which are provisions held by the project sponsor for possible changes in project scope or quality that can be used to mitigate cost and/or schedule risk. Always plan ahead, be

TABLE 12.2: Risk Mitigation Strategies for Technical, Cost, and Schedule Risks

TECHNICAL RISKS	COST RISKS	SCHEDULE RISKS
Emphasize team support and avoid stand-alone project structure	Increase the frequency of project monitoring	Increase the frequency of project monitoring
Increase project manager authority	Use WBS and PERT/CPM	Use WBS and PERT/CPM
Improve problem handling and communication	Improve communication, project goals understanding, and team support	Select the most experienced project manager
Increase the frequency of project monitoring	Increase project manager authority	
Use WBS and PERT/CPM		

proactive in risk management. Questions that need to be addressed in a Risk Management Plan include:

1. Why is it important to take or not take this risk in relation to the project objectives?
2. What specifically is the risk and what are the risk mitigation deliverables?
3. How is the risk going to be mitigated?
4. What risk mitigation approach should be used?
5. Who are the individuals responsible for implementing the risk management plan?
6. When will the milestones associated with the mitigation approach occur?
7. How much resources are required to mitigate risk [5], [6] ?

12.7 RISK RESPONSE CONTROL

Risk response control involves executing the risk management processes and the risk management plan to respond to risk events. Risks must be monitored based on defined milestones and decisions made regarding risks and mitigation strategies. If there are no contingency plans in place, then workarounds or unplanned responses to risk events are necessary. A tool for maintaining an awareness of risk throughout the life of a project is the tracking of the project's Top 10 risk items. The list of the Top 10 project risk items should be reviewed periodically and modified as the risk ranks change. Hence, a listing of the current ranking, previous ranking, number of times the risk appears on the list over a period of time, and a summary of progress made in resolving the risk item should be developed. Project managers may use Project Risk Management software (databases or spreadsheets) to assist in keeping track of risks and quantifying risks. There are several more sophisticated risk management software available in the market that may help the project manager in developing models and/or simulations to analyze and respond to various project risks.

12.8   SUMMARY

The Risk Management Processes include identifying the risk or hazard, assessing or analyzing all risks and/or hazards, making risk response decisions, implementing controls to mitigate the risk, and supervising or overseeing the implementation of the responses or corrective actions. Unlike crisis management, good project risk management often goes unnoticed, because well-run projects appear to be almost effortless, but, in reality, a lot of work has gone into running the project well. Hence, project managers should strive to make their jobs look easy; thus, reflecting the results of well-run projects.

REFERENCES

[1] Risk Management. Wikipedia. Encyclopedia. Available: http://en.wikipedia.org, 2005.

[2] D. Faherty, "U.S. Navy, operations risk management," presented at the 2nd Workshop on Risk Analysis and Safety Performance Measurements in Aviation, FAA, Atlantic City, NJ, Aug. 2000.

[3] RISK + C/S Solutions Newsletter. Available: http://www.cssi.com/

[4] M. S. Dorfman, *Introduction to Risk Management and Insurance*, 6th ed. Englewood Cliffs, NJ: Prentice-Hall, 1997, ISBN 0-13-752106-5.

[5] B. C. Chadbourne, "To the heart of risk management: Teaching project teams to combat risk," in *Proc. 30th Annu. Project Manage. Inst. Semin. Symp.*, Philadelphia, PA, Oct. 10–16, 1999.

[6] A. Jaafari, "Management of risks, uncertainties, and opportunities on projects: Time for a fundamental shift," *Int. J. Project Manage.*, vol. 19, no. 2, pp. 89–101, Feb. 2001.

CHAPTER 13

Project Closeout

Project Closeout is the last major stage of a project's life cycle and is completed when all defined project tasks and milestones have been completed, and the customer has accepted the project's deliverables.

So, what is involved in closing projects [1]? Project Closeout includes the following actions:

1. First and foremost is the gaining of stakeholder acceptance of the final product,
2. Verification of formal acceptance by stakeholders and steering committee,
3. Bringing the project and all its phases to an orderly end,
4. Verify that all of the deliverables have been completed and delivered,
5. Completing a Project Audit (usually internal audit),
6. Redistributing resources; i.e., staff, facilities, equipment, and automated systems,
7. Closing out all financial issues such as labor charge codes and contract closure,
8. Documenting the successes, problems, and issues experienced during the project,
9. Documenting "Lessons Learned,"
10. Producing an Outcomes Assessment Report,
11. Completing, collecting, and archiving Project Records,
12. And finally, it is recommended that the Project Team celebrate project success.

13.1 CLOSING PROCESSES AND OUTPUTS

Most of the closing processes involve the communications and procurement knowledge areas as shown in Fig. 13.1.

The process of "Administrative Closure" involves collecting project records, verifying and documenting project results to formalize acceptance of the products produced, analyzing whether the project was successful and effective, ensuring products meet specifications, and archiving project information for future use. Table 13.1 Column 3, row 2, shows the outputs that are the result of the administrative closure.

TABLE 13.1: Closing Processes and Outputs

KNOWLEDGE AREA	PROCESS	OUTPUTS
Communications	Administrative closure	1. Project archives 2. Formal acceptance 3. Lessons learned
Procurement	Contract Close-out	1. Contract files 2. Formal acceptance 3. Formal closure

13.1.1 Administrative Closure

The issue of primary importance with project closure is the acceptance of the product or project deliverables by the customer for which they were created [2]. During the administrative closure, the project manager should conduct a formal "Final Acceptance Meeting." The best way is to convene a final meeting with stakeholders to review the product delivered against the baseline requirements and specifications. It is always a good policy to make the stakeholders aware of the baseline deviations, with justifications for the deviations, and of future action plans to correct or to waive the deviations. Deviations from the established baseline should be documented and approved at the committee for subsequent signatures by responsible executive managers of the organization. Open Action Items or program level issues can be officially closed or reassigned to the support organization. Drawing the stakeholders together in a single meeting helps avoid clearing up open issues on an individual basis.

13.1.2 Approval Verification

Approval is verified via the signature of a project closure document by the stakeholders who signed the original project baseline documentation (i.e., the Project Plan). Acceptance document should be customized to the particular project to include:

1. Pertinent deliverables,
2. Key features, and
3. Information about final product delivery.

13.1.3 Procurement Contract Closure

Contract closure is the process of terminating contracts with outside organizations or businesses. Contracts may be for providing technical support, consulting, or any services. Contracts are usually brought to closure upon contract completion, early termination for cause, such as, failure to perform. Closing a contract usually requires assistance from the Contracts Administrator, since close attention must be paid to ensure all obligations of the contract have been met or formally waived and to prevent any liability for the organization. Normally, procurement will conduct the "Final Contract Review Meeting." Project managers should make a checklist of all items that must be addressed during contract closure, such as:

1. Review contract and related documents,
2. Validate that the contractor has met all of its contractual requirements,
3. Document any contractor variances,
4. Resolve contractor variances and issues,
5. Validate that the organization has met all of its contractual requirements,
6. Document organization's variances and issues,
7. Resolve Agency organization's variances
8. Ensure that all vendor responsibilities have been transferred to the organization or another vendor,
9. Terminate current contract, and
10. Verify that all contractual obligations have been met or formally waived.

13.2   OUTCOMES ASSESSMENT MEETING

"Another meeting," you ask? Well, so far, only two have been covered. Do not be surprised if there are more. In conducting "Outcomes Assessment Meetings," project managers provide a forum for discussing the various aspects of the project with the primary focus on project successes, problems, and issues, "Lessons Learned," and recommendations for future process improvements. Program managers should use the information and documentation from the "Final System Acceptance Meeting" as a basis for the Outcomes Assessment Meeting discussions. Outcomes Assessment Meetings are usually attended by the project manager as chairman or moderator, all members of the Project Team, along with representation from Stakeholders, Executive Management, Maintenance, and Operations Staff [3]. It is always wise to include some oversight members that are external to the project and even the organization.

Typical questions that should be addressed in the Outcomes Assessment meeting include the following:

1. To what extent did the delivered product meet the specified requirements and goals of the project?
2. Was the customer satisfied with the end product?

3. Were cost budgets met?

4. Was the schedule met?

5. Were risks identified and mitigated?

6. Did the project management methodology work?

7. What could be done to improve the process?

13.3   OUTCOMES ASSESSMENT REPORT

After the meeting, the project manager and project team must generate an Outcomes Assessment Report that documents the successes and failures of the project. The Outcomes Assessment Report provides an historical record of the planned and actual budget and schedule. The report should include description with rationale of selected metrics that were collected on the project and were based on documented procedures. The report should also contain recommendations for future projects of similar size and scope [4]. Outcome Assessment Reports should contain the following information:

1. Project sign-off,

2. Staffing and skills,

3. Project organizational structure,

4. Schedule management,

5. Cost management,

6. Risk management,

7. Quality management,

8. Configuration management,

9. Customer expectations management,

10. Lessons learned, and

11. Recommendations for process improvement.

13.4   TRANSITION PLANNING

Before projects are closed, it is important for organizations to plan for and execute a smooth transition of the project into the normal operations of the company. Most projects produce results (resources) that are integrated into the existing organizational structure, some may require modification of the organizational structures, whereas, some other projects are terminated before completion. Additionally, the organization must develop a plan on how to redistribute resources; i.e., project team members, support staff, materials, facilities, equipment, and automated systems, before projects are closed or cancelled. The Project Manager is responsible for turning over to the operations and maintenance organizations all documentation that has anything to do with the product including design documents, schematics, and technical manuals.

13.5 PROJECT DOCUMENTS TO BE ARCHIVED

Some of the typical project documents to be archived include:

1. Project Business Case,
2. Project Plan, including Project Charter, Project Scope Statement, Risk Assessment, Risk Mitigation,
3. Management Plan, Communications Plan, Quality Assurance Plan, etc.,
4. Financial Records,
5. All correspondence on project matters,
6. Meeting Notes,
7. Status/Progress Reports,
8. Procurements and Contract File,
9. Test Plans and Results,
10. Technical Documents,
11. Files, Programs, Tools, etc., placed under Configuration Management, and
12. All other documents and information pertaining to the project.

13.6 CRITICAL SUCCESS FACTORS

The most critical factors used to measure project closeout success are first and foremost acceptance by the end-user, followed by having achieved the business objectives and anticipated benefits. Next factors are the achievement of project objectives and knowledge transfer. The final factor is archiving of all project materials.

13.7 SUMMARY

Generally, Project Closeouts include the following key elements:

1. Verification of formal acceptance by Stakeholders and Steering Committee,
2. Redistributing resources; i.e., staff, facilities, equipment, and automated systems,
3. Closing out any financial issues such as labor charge codes and contract closure,
4. Documenting the successes, problems, and issues of the project,
5. Documenting "Lessons Learned,"

6. Producing an Outcomes Assessment Report,

7. Completing, collecting, and archiving project records, and

8. Celebrating Project Success: "End it with a Bang!"

REFERENCES

[1] Project Close Out Check List. Project Management. University of Washington, Seattle, WA. [Online]. Available: http://www.washington.edu/computing/pm/end/closeout.html, 2002.

[2] Close the Project. Project Management. University of Washington, Seattle, WA. [Online]. Available: http://www.washington.edu/computing/pm/end, 2002.

[3] Close/Audit Phase. Trainers Direct. [Online]. Available: http://www.trainersdirect.com/resources/Project%20Management/CloseoutPhase.htm, 2005.

[4] Project Closeout Report. Document 06-114. (2006, Apr.). History. Texas Department of Information Resources, Austin, TX. [Online]. Available: http://www.dir.state.tx.us/pubs/pfr/06-114/instruction.pdf.

CHAPTER 14

Project Design Reviews

Part of this chapter is based on excerpts from a slide presentation given at the Naval Air Warfare Center in 2000. Why should companies conduct design review? The main reason may be that design reviews are required by some "Regulation" in all government departments or agencies dealing with commercial or military products, e.g., Food and Drug Administration, Federal Aviation Administration, Department of Commerce, National Institute of Standards and Technology, or Department of Defense Regulations. Following the regulation, guidelines are of interest to those companies or industries that propose to enter the U.S. commercial market. Not following the regulations could result in product "Liability" issues. Most court rulings are based on the engineering practice of following "Good Common Practice" (standards and regulations) and abiding by professional ethical codes that hold the "Health and Welfare of the Public" as paramount. The purpose of including this chapter is to provide students and new employees the guidance necessary in the preparation and conduct of Preliminary Design Reviews (PDR) and Critical Design Reviews (CDR).

14.1 PRELUDE TO CONDUCTING A DESIGN REVIEW MEETING

The primary objective in conducting design review processes is to ensure that the design fulfills the performance requirements. In conducting design reviews, the program manager or his designated chairperson must first identify the design review objectives, list the entry and exit requirements for design reviews, and state the responsibilities for the committee conducting design reviews. The Design Review Committee Chairperson must form the committee with project team members, stakeholders, sponsors, technical area experts, and independent members; and coordinate availability to ensure participation by essential members in the Design Review. Additionally, a "Meeting Agenda" is prepared, coordinated with the committee members, and subsequently accepted prior to the meeting. Before entering into the design review, the chairperson must ensure that committee members have necessary documents, such as:

1. Requirements Traceability to Specifications Matrix, which is sufficient for the preliminary design review.

2. Math Model Report, which is sufficient for the preliminary design review.

3.   Any design documents; i.e., AutoCAD drawings, circuit drawings, or layouts.

4.   Risk assessments and risk mitigation plans that are to be formally addressed.

When asked, "Is there a format or any one format for the design review?" The response is, "Not Really, because design reviews are carried out at various intervals (phases) during the development of a product." Hence, the design reviews may address specific points or all of the major concerns in a phase. Table 14.1 contains an example of a design review agenda. During a test phase prior to any tests being conducted, the project manager may conduct a review of the test plan, testing procedures, and verify availability of test personnel and that all necessary test equipment are in place, functioning, and calibrated to some standards lab with current calibration stickers. After the tests are conducted and analyzed, the project manager would conduct another review of the test results, listing all deficiencies, and perhaps formulate options for correction of the deficiencies.

14.2   ENTRY CRITERIA FOR DESIGN REVIEW

Entry Criteria are the minimum essential items necessary to enter into a design review. If the design review is not a preliminary review, then one of the most important criteria for entry is that there are "No outstanding prereview action items." The project manager or the review committee chairperson should define the design baseline and provide the framework for the design review; including specific items in the Breakout of Tasks Work (WBS), requirements traceability matrix, specifications, and items describing the design. Prior to the design review meeting, the chairperson should submit the meeting agenda with specified items to members for review giving ample time for members to make corrections or comments.

The Requirements Traceability Matrix (RTM) provides the information-linking requirements to all of the design documentation, and is the tool that enables a company to verify that all of the design requirements are being addressed. If the product includes software, then the Software Design Documentation should adequately disclose software design approach information. Typical software design documents include:

1.   Math calculations and Model simulation reports,

2.   Software Design Description, requirements, and specifications,

3.   Interface Design Description, requirements/specifications, and

4.   Database Design Description,

5.   Software Test Plans, and

6.   All Software Development Folders (flow charts, source code listings, etc.).

TABLE 14.1: Example of a Design Review Agenda (Preliminary Design Review) ©2002 DRM Associates

REVIEW TOPIC	REQUIRED OUTPUTS
Project Definition	Program/Team Charter
• Customer changes to the program since last review (if any)	Program Requirements/Deliverables
	Budget & Schedule Changes
Concept Approach Changes Since Last Review (if any)	Product Specifications
• Changes to customer requirements & specifications since last review (if any)	Concept Design
• Specification issues (if any)	
• Changes to system architecture and concept approach for the system (if any)	
• Changes to product concept design (if any)	
Product Design	Component Drawings/CAD Models
• Review of design concept for each product	Assembly Drawings/CAD Models
• Product structure walk-through	Schematic/Net List
• Schematic and functional design review (if applicable)	
• Assembly drawing/model review	PCB Layout
• Component drawing/model review	
• Part specifications, significant characteristics, and tolerances	Product Bill of Material
• Design for manufacturability, design for assembly, and mistake-proofing review	
• Design and drawing/modeling standards compliance	
• Technical issues and risks	

(*continued*)

TABLE 14.1: (*continued*)

REVIEW TOPIC	REQUIRED OUTPUTS
Product Testing and Verification • Test requirements and plan • Test results • Issues in meeting specifications	Test Requirements Test Plan Test Results
Preliminary Process Design • Process approach and operation flow • Feedback from engineering model build • Tooling, fixture, production equipment, and test equipment requirements • Tooling and fixture design • Tooling, fixture, production equipment, and test equipment cost estimates/quotes	Build Report Process Flow Diagram Tooling and Fixture Design Tooling & Fixture Cost Estimates
Quality Planning • Quality issues on similar components and countermeasures taken • Design FMEA, reliability issues, and mitigation steps • Process FMEA, reliability issues, and mitigation steps • Control Plan for design validation build	Design FMEA Process FMEA Control Plan
Supplier Sourcing and Status • Supplier selection and capability for each component • Production, capability, quality, lead-time, and cost issues for each supplier and component • Inbound packaging requirements or specifications	Supplier Plan

TABLE 14.1: (*continued*)

REVIEW TOPIC	REQUIRED OUTPUTS
Product Cost and Profitability • Current cost estimate compared with target cost • Product profitability	Target Cost Worksheet
Program Plan and Management • Project plan • Schedule issues • Resource issues • Process deviations	Project Plan Project Budget
Review of Issues and Follow-up Actions	Open Issues List

Permissions to use the tables were granted by DRM Associates, November 8, 2006. Note the tables were primarily designed for the automotive industry; however, the tables contain general information that can be tailored to specific design reviews.

Hardware Design Documentation should adequately disclose all hardware design information. Typical hardware design documents include:

1. Theoretical math calculations, Models, and/or Simulation Reports,
2. Hardware requirements and specification,
3. Hardware Interface requirements and specification,
4. Circuit and/or AutoCAD drawings with parts listings,
5. Hardware Test Plans and Test Results Documents, and
6. Training Plan and Training Manuals.
7. Additionally, Risk Assessments and Mitigation Plans must be formally addressed within the design documentation.

14.3   CONDUCTING THE DESIGN REVIEW

Normally, the Design Review Chairperson is responsible for conducting the design review; however, for student design, the Team Chairperson is responsible for conducting the design review. In most organizations, it is essential that customer representatives and users participate in design reviews; however, for student teams, it is essential that the industry sponsors participate in the review. For either commercial companies or student teams, it is essential to include a representation from appropriate specialties; e.g., hardware, systems, software, human factors integration, facilities engineering, etc. It is recommended that student teams hold their review at the sponsor's facility. Chairpersons should make sure that all participants have ample opportunity to address questions, issues, and concerns. Typically, the design reviews take the form of formal presentations by the design team to the full Review Committee. The presentations should begin with a brief overview of the overall program (scope, deliverables, and milestone schedules) to set the stage for the design briefs and an overall systems perspective reflecting the major subsystems and how they interface to comprise the total system. The briefing format should reveal the following:

1. Identification of each requirement referenced to the appropriate specification paragraph and/or work task.

2. The design approach for preliminary design review (PDR) or detailed design review (CDR) for each requirement. Illustrations should be included wherever feasible.

3. Risk assessment for each requirement and the risk mitigation techniques employed to manage the risk.

4. Risk management plan including who is responsible for carrying out the risk mitigation strategies.

5. Safety and Human factors.

14.4   DESIGN REVIEW OUTPUT

At the conclusion of the design review, minutes of the meeting or a Summary Report of the meeting must be generated. Documents to be submitted with the Summary Report include all the design documents used during the course of the design review. As a minimum, the following documents should be attached or forwarded with the report:

1. Contractor's Proposal or copy of signed contract,

2. SOW (Tasking with Gantt Chart),

3. Specifications,

4. Requirements Traceability Matrix (RTM),

5. All Circuit, AutoCAD Drawings,

6. All Design Documents, and

7. Requests for Action (RFA).

Requests for Action (RFA) are formal forms generated to document questions, issues, and concerns that surface during the design review. It is essential that suspense dates and responsibilities for resolving the RFAs be assigned before completion of the design review. The student design teams must be provided some form of RFA as shown in the Appendix of the Design Review Report.

14.5 EXIT CRITERIA

Exit criteria are the minimum essential items necessary to successfully complete a design review before proceeding into the next phase. Therefore, project managers should review items specified in the statement of work, the specifications, and the requisite items describing the design have been successful resolved and all action items are closed. They should also ensure acceptance of required items and acceptance of the design review minutes. Is it over now? It is not over until management has made the determination from the Exit Documents as to whether or not the program/project/design is ready to proceed into the next phase based upon successful completion of the exit criteria (Signatures are required).

REFERENCES

[1]   *Technical Design Reviews*, Naval Air Warfare Center, Training Systems Division, 2002.

[2]   DRM Associates. (2002). *Example of a Design Review Agenda (Preliminary Design Review)* [Online]. Available: http://www.npd-solutions.com/designreview.html

CHAPTER 15

Making Technical Decisions

This chapter is added to this book on project management to help student teams learn to make rational informed decisions during their senior design projects. Even though everyone makes daily decisions, not many of those decisions are associated with project management and technical matters. The psychology of decision-making varies among individuals. Comedians poke fun at the decision process between men and women when they draw analogies to shopping differences. Women spend hours in a store buying one item, because they search for all sorts of alternatives with the lowest price being one of the factors; whereas, men go in, see what they came for, get it, and they are out of there in a couple of minutes. Comedians propose that the basis for the difference in decision behavioral patterns goes way back to our cave dwelling ancestors when women would go "berry picking" and they were very picky about their berries. This one is good, no good, ok, bad; hence cavewomen would spend hours picking berries. Cave men, on the other hand, were hunters. "There is the rabbit, kill it!" Off goes the arrow; "got it, now time to go home and eat! Ugh!" Truly, this caveman approach is not the way of modern decision-making with today's technological advances.

Webster's dictionary [1] defines "Decision" as the act of making up one's mind; the result or conclusion arrived at by deciding. "Decision-Making" is defined in Webster's dictionary as the process by which decisions are made. The Center for the Study of Work Teams (CSWT) at the University of North Texas [2] defined "Group Decision-Making" as the process of arriving at a judgment based upon information and the feedback of multiple individuals.

15.1 GROUP DECISION-MAKING PROCESS

Various organizations use different Decision-Making Models to establish a systematic means of developing effective group decision-making. Since a multiplicity of models exists, only the four basic "Group Decision-Making Models" will be discussed:

1. Rational Model,
2. Political Model,
3. Process Model, and
4. Garbage Can Model.

15.1.1 The Rational Model

The Rational Model is based on an economic view of decision-making and grounded on goals/objectives, alternatives, consequences, benefits or opportunities, and optimization.

The Rational Model assumes that complete information regarding the decision to be made is available; thus, decision-makers consistently assess advantages and disadvantages of alternatives with goals and objectives in mind. Additionally, they will evaluate the consequences of selecting or not selecting each alternative. Finally, the alternative that provides the *maximum utility* (i.e., the optimal choice) will be selected as the best choice or solution. The rational model is often used as the baseline against which other models are compared. With the Rational Model, decisions are made deductively by determining goals and objectives to be obtained, evaluating the potential alternatives based on the information at hand and choosing the optimal alternative.

The advantage of the Rational Model is that it uses a logical and sequential approach. The disadvantage of the model is that it assumes no intrinsic biases to the decision-making process.

15.1.2 The Political Model

With the Political Model, groups or individuals do not decide through rational choice with regard to objectives; instead, the decision-makers are motivated by and act on their own needs and perceptions. The model involves bargaining among the decision-makers as each one tries to get his/her perspective to be the one of choice and does not involve or require making full information available. The only advantage of the Political Model is that it emulates how the real world operates (i.e., bargaining related to personal agendas). The greatest disadvantage of the model is that the best solution or decision may not be selected; for example, decision-making in the "U.S. CONGRESS" is based along party or constituency lines rather than on rational goal-oriented or technical merits [2].

15.1.3 The Process Model

Process Model decisions are based on standard operating procedures, or pre-established guidelines within the organization in which conformity to past and present organizational rules and standards is an integral part. Conformity relates to the fact that reasoning for the decision is based on the predetermined guidelines: REGULATIONS! Again, large government organizations too often tend to quote regulations [2].

15.1.4 The Garbage Can Model

The last decision model is the Garbage Can Model, which is used to make judgment or decisions on tasks within organizations where the technologies are not clear. In the Garbage Can Model, the involvement of participants as well as the amount of time and effort given to the decision process fluctuates such that choices are usually inconsistent and not well defined. However, the model provides a real-world representation of the nonrational manner in which decisions are often made by individuals and within some organizations. "Ad Hoc" decisions made by "Flying by the seat of the pants!" are not the most efficient means of making a decision. Let us hope that student design teams avoid and never use this model in making technical (engineering) decisions [2].

15.2   U.S. NAVY EXECUTIVE DECISION-MAKING FRAMEWORK

The U.S. Navy includes definition, analysis, decision, reconciliation, and execution phases in their Decision-Making Framework [3]. The definition phase requires describing in detail the following:

1. Problem Statement,
2. Decision Objectives,
3. Context,
4. Boundaries or Limits, and
5. Analytic Objectives.

The analysis phase requires development of decision criteria based on examining the validity, reliability, practicality of the solution, the uncertainty and risks, the analytical method, sensitivity analysis, the decision model, and alternatives.

The decision phase requires taking time to review the entire decision process for timing, any spillover effects, organizational issues, political issues, evaluating the internal decision, and performing a reality check on the process and solution.

The reconciliation phase may be thought of as "Smoothing ruffled feathers," that is removing all negative effects on participants. One may also consider this phase as conflict resolution among participants. What strategies or techniques to use in conflict resolution, i.e., win–win compromise with mutual gains or zero-sum on the scorecards?

The last phase is execution of the decision. Implementation of any decision or solution requires planning on how to carry out the decision and verifying that the decision is being carried out correctly. The plan must detail who (Which individual or organization is responsible for implementing the decision?), how (How is the decision to be executed?), and what controls? The execution of the decision is not simply, "Here is a memo, go do it!" Implementation of the decision should be "verified" by those making and issuing the implementation plans or directives. Verification requires measurement of some metric, some feedback mechanism on the progress; i.e., EVA, QA, etc., and a mechanism for adjustments to the implementation.

15.3   DECISION MATRIX OR UTILITY FUNCTION

Lessard [4] and Bronzino [5] used a method for making technical decisions based on the Rational Model, but is referred to in various terms, i.e., Decision Matrix [6], Weighted Function Evaluation [5], and Utility Function [4].

15.3.1 Weighted Function Evaluation

Weighted function evaluation requires assigning a weighing factor to denote the relative importance of each of the desired attributes; usually, 0–10. In collaboration with the other evaluators, it determines how well each device meets each attribute with individually assigned scores (between 0 and 10), and then multiplies the scores by the respective weighing factor to determine a weighted score for each attribute for each device. The total weighted scores (Averaged for multievaluators.) are used to determine the overall rating for each device. The technical decision uses these ratings to determine the relative ability of each device in meeting the specified requirements [4].

15.3.2 Authors' Recommendations

The authors recommend that student teams use the "Utility Function Method" for evaluation and decision-making. Teams should start by evaluating product objectives and user requirements, and translate requirements into engineering system or device specifications. Lastly, apply the "Utility Function" in making technical or engineering decisions.

15.3.3 Utility Function

The Utility Function is a quantitative methodology appropriate to assess the relative merits of the available methods, systems, or devices. The first step is to determine essential variables and their respective weights. Variables are defined as those factors necessary to evaluate with some figure of merit the most useful system. A limitation of the utility function analysis is that the outcome of the evaluation may not be the same for other applications, e.g., purchasing a medical device for use in a medical evacuation aircraft or a large well-equipped hospital. Some consider the fact that the function may neither be unique nor does the function include all possible variables as a limitation. The fact is that very seldom are all the critical factors known and included.

The selection of factors is a commonsense approach in which it is necessary to evaluate the importance of what is being measured; e.g., in the determination of vital life signs. How the measurement is obtained is not as important as the need to evaluate how accurate the system may measure an essential vital sign. The next step is to prioritize or assign Weighting Factors by order of importance.

15.4   FACTOR WEIGHTS

Factor weights are the coefficients or multiplying factors by which the variables (factors) are multiplied. The magnitudes or values assigned to the factor weights are not unique. One method of obtaining the weights is to conduct a survey by having a large number of engineers or specialists in the area assign values based on some guidelines and criteria. Surveys require an extended period of time to collect and analyze. One should always question, "How dependable and reliable will the final results be?" The answer is that the results will depend on how well the selection of factors and assignment of weights describe or represent the usefulness of the criteria for the specified conditions.

15.5   GRADING SCALE

The next step is to select a scale for the factor weights, for example, one may use a scale from 5 to 0 in three discrete levels to represent:

1. Most desirable (5 points),
2. Acceptable (3 or 2 points), or
3. Unacceptable (1 or 0 points).

A scale may be doubled for the very important factors; thus, the factors were given weights based on their relative importance; for example: It is determined that the types of electrodes used are very important and are weighted 10 points with three levels and points:

1. Noncontact—10 points.
2. Noninvasive without media—5 points.
3. Noninvasive with media (i.e., gel)—0 point.

In evaluating, the relative merit and/or utility value of each candidate system is calculated after considering all the factors in the model [Eq. (15.1)]:

$$y_n = \sum a_{ij} x_i \qquad (15.1)$$

where

x_i is the ith factor,
a_{ij} is the weight of the ith factor with j degrees, and
y_n is the utility value of the nth system.

Since the factors are descriptors, $a_{ij} x_i$ is not a product, but rather a designation of points that are summed to yield a utility value.

15.6   SUMMARY

In making technical decisions, teams should evaluate objectives and user or system requirements. Then, convert requirements into engineering specifications of a system or device. Be sure to use the Utility Function Method for evaluation of alternatives leading to a rational, systematic "Group Decision" based on the average of individual analyses.

REFERENCES

[1] *Webster's New Collegiate Dictionary.* Springfield, MA: G. & C. Merriam, 1973.

[2] Center for the Study of Work Teams, *Group Decision Making Within the Organization: Can Models Help?* Denton, TX: University of North Texas, 1996.

[3] US Navy Command and Staff School. (1987). Navy Web Site. [Online].

[4] C. S. Lessard and W. C. Wong, "Evaluation of noninvasive measurement methods and systems for application in vital signs detection," USAFSAM-TR-85-44-PT-1, 1983.

[5] J. Bronzino, *Management of Medical Technology.* London, U.K.: Butterworth, 1992.

[6] *Executive Decision.* New York: McGraw-Hill, 1964

CHAPTER 16

Management of Team Conflict

This chapter is presented in part from an article by Vern R. Johnson in the IEEE-USA Today's Engineer with the permission of the IEEE-USA Publishing [1].

Conflict is defined in Webster's dictionary [2] as a noun meaning argument, dispute, or controversy; and, conflict is also defined as a verb meaning to disagree, to differ, to vary, or to disturb.

Disagreements and conflicts are often described as the heat generated by change. Some individuals claim they can "feel" when conflict exists; whereas, others believe the conflicts "...are unavoidable." Nevertheless, "A Conflict will arise!" whenever individuals disagree in a heated verbal exchange about what direction to take when changes occur or are necessary. Usually, changes due to business directions, customers, and technology within a project may require realigning strategies and goals around the new direction with team members' agreement or compromise about the realignment [1].

Three basic causes of conflict include [1]:

1. *Information*: Refers to the conflict that is based on incorrect information or the lack of information. For example, if someone does not know when or where the team is to meet and does not attend the meeting, that individual's absence will limit the team's progress and may create "Conflict" between members.

2. *Goals/Roles*: Refers to the conflict that will arise if some team members do not understand the team's task, or if members do not know specifically what their assignments are, those members cannot align themselves with the project or the team.

3. *Values*: Refers to the conflict that will be present if team members do not share values relative to the task and the approach used in the tasks.

The ability of individuals and/or teams to accomplish tasks has been correlated directly with the team's ability to handle conflict. Since it is well known that conflict interferes with team productivity, when a team experiences conflict, it is essential that its members resolve the conflict before moving forward [1].

TABLE 16.1: Communication Model [1]

	INDIVIDUAL A	INDIVIDUAL B
1	Data	Listen
2	Interpret	Listen
3	Feelings	Listen
4	Needed action	Listen
5	Listen	Echo
6	Listen	Decide

Often, what appears to be a conflict may simply be a misunderstanding. One approach to determine if a conflict exists is to use the communication model in Table 16.1.

Individual A does the following four things while Individual B listens [1]:

1. Give the data as an objective statement, without making a judgment or offering feelings. For example, "I see that you . . . " or "I noticed that . . . ."

2. Make an interpretation. Individual A shares his/her judgment of what the data means. "I interpret this to mean . . . ."

3. Identify the feelings that result from interpretation. There are always "feelings"; i.e., variations of anger, sorrow, joy, or fear associated with conflict. The individual A should make a simple statement that recognizes those feelings, such as, "I feel . . . about it" or ". . . and it makes me angry."

4. State the need to be filled or the required action. For example, "I would like you to . . . ," or "I want you to . . . as a demonstration that you are still part of the team."

    Then, individual A should stop talking and listen while individual B responds (steps 5 and 6).

5. Echo the expected action. Individual B should parrot back what was just heard to validate understanding of the message. For example, "I see that you are angry and you don't think I care," or "You want me to . . . as a demonstration that I am part of the team."

6. Decide what you are willing to do about person A's concern and respond accordingly; e.g., "To prove myself, I will . . . ," or "I had no idea you would interpret my actions that way. In the future, I will . . . ."

If individual B's answer is, "I will cooperate with your request," then the situation resolves itself. However, if individual B's answer is, "I will ignore your need or requested action," then there is a conflict that must be resolved before individuals A and B can move forward and be productive.

Without a doubt, most individuals prefer to give advice or feedback, than to take or receive feedback. However, sometimes it may be necessary to notify team members that some of their actions are not acceptable, often resulting from oversight regarding ground rules, shared responsibilities, or personal behaviors. Being invited to examine your behavior is receiving a "wake-up call" from one's peers that may be an emotional, awkward, or uninvited experience. It is critical that the person who approaches a team member to give appropriate feedback knows how to give feedback without antagonizing the team member or members. Therefore, it is important for both the individual giving feedback and the individual receiving feedback to express appropriate attitudes during any discussion [1].

16.1   GIVING FEEDBACK

In giving feedback, make the meeting cordial and individuals should give appropriate feedback in a simple, straightforward way, but with some degree of care. It is not a time for team members to gang up on the person or to express a litany of concerns. However, do not withhold concerns until they have become overwhelming, because the sooner the problems are approached, the easier it will be to resolve them. The focus should be on the individual's behaviors and the results of those behaviors, or on the technical merits of an approach, rather than on the individual's personality. Cite a specific situation as an example of unacceptable behavior and describe the change that may need to be made; however, when expressing concerns, avoid giving advice and then allow time for the individual to respond [1].

It is just as important for people receiving corrective feedback to know how to respond. Here are some basic guidelines for receiving feedback. Since there is a problem and one needs to understand the basic problem, therefore, listen with an open mind. Make sure to understand what teammates are saying, therefore, do not hesitate to ask questions for clarification if necessary. Do not overreact, or agree with, or reject the confrontation, the initial objective this time is to gather information. Express appreciation for the information, since teammates are taking a risk by trying to help [1].

16.2   CONFLICT RESOLUTION METHODS

There are four conflict resolution methods [1]:

1. *Avoidance*: Avoidance is the general method used when it has been determined that the relationship is not important enough to save. For example, "I can't handle this. I'm

out of here." or, "The cost of complying with your request is just too high. Let's call the whole thing off."

2. *Exercise power*: When power is exercised, an individual takes an assertive position based on power or position; e.g., "I am in control." or "I am the boss."

3. *Who is right?* This method requires a third party to mediate the conflict. If the conflict is based on flawed information or confusion over goals or roles, it may help to go to an expert or consult a reference book to find an acceptable solution.

4. *Interest-based.* When a conflict exists between individuals, each takes a position, they anchor themselves to their positions, and they become entrenched with a barrier between them.

What should a project manager do when there appear to be conflicts among team members? First, find out "why" the individuals took the position they did. What is behind it? From the answers, determine what interest the individuals have in common. The objective is to have the individuals concentrate on the common interest rather than on their differences. Common interests can lead to compromise, which, in turn, helps those in conflict to relax from their entrenched positions [1].

16.3   SUMMARY

A simple three-step process that should be used by the project manager includes:

1. Achieve contact
   a. Validate the feelings of other people.
   b. Learn why they have taken a position.
   c. Understand them.

2. Boil down the problem
   a. Ask clarifying questions about the issues that appear to exist.
   b. Prioritize these issues.

3. Choice making
   a. Attempt to identify alternatives that can be chosen to provide an appropriate compromise.
   b. Protect the common interest.

All members must take responsibility for implementation, and once implemented, the conflict will recede. If accountability for implementation is not verified, the conflict can return without warning. As a final note, after a conflict is resolved, it is amazing how effective a *"Thank You"* is at bringing goodwill back into the relationship [1].

REFERENCES

[1]   V. R. Johnson. (2005, Jan.) Managing conflict in a small team setting. *IEEE-USA Today's Eng.* [Online]. Copyright IEEE 2006. Available: http://www.todaysengineer. org/2005/Jan/conflict.asp.

[2]   *Webster's New Collegiate Dictionary.* Springfield, MA: G. & C. Merriam, 1973.

Author Biography

Charles S. Lessard, Ph.D., Lt Colonel, United States Air Force (Retired), is an Associate Professor in the Department of Biomedical Engineering at Texas A&M University. His areas of specialization include Physiological Signal Processing, Design of Virtual Medical Instrumentation, Noninvasive Physiological Measurements, Vital Signs, Nystagmus, Sleep & Performance Decrement, Spatial Disorientation, G-induced Loss of Consciousness G-LOC). Neural Network Analysis. Dr. Lessard received a B.S. in Electrical Engineering from Texas A&M (1958) a M.S. from the U.S. Air Force Institute of Technology (1965), and Ph.D. from Marquette University (1972).

As an officer in the U.S. Air Force, Lessard was a pilot of F86L Interceptors and B-52G Strategic Bombers. He also served as Research Scientist and Chief of Biomedical Engineering Research for the Aerospace Medical Division of the School of Aerospace Medicine, at Brooks Air Force Base, Texas. In this capacity he planned and directed efforts in biomedical projects associated with the Manned Orbiting Laboratory Program (MOL), developed medical instrumentation (EEG Analyzer), conducted research on computer on the analysis of sleep brainwaves and cardiac signals, and the effects of zero-gravity (0-G) on the cardiac response during valsalva maneuvers. U.S. Air Force Medical Research Laboratories, Wright-Patterson AFB, Lessard with Biocybernetics Wing Engineering and worked on neural networks, self-organizing controls (SOC), and remotely piloted vehicles. He was the Field Office Director. Program Manager, with the Electronics Systems Division of the Air Force Systems Command during the installation and testing of Spain's Automated Air Defense System as a part of the Treaty of Friendship and Cooperation between the US and Spain. Dr. Lessard retired from the U.S. Air Force in 1981 after serving as the Deputy Director Bioengineering and Biodynamic Division at Aerospace Medical Research Laboratory (AMRL), Wright-Patterson Air Forces. He began his academic career with Texas A&M University in 1981. His program management experiences are applied in his two Senior Design Courses.

Charles Lessard was a Senior Scientist for Veridian Inc. at Brooks Air Force and lead scientist for KRUG Life Sciences, Inc. in the psychological and neurophysiological manifestations of spatial orientation, mechanisms of spatial orientation in and countermeasures against spatial disorientation. Additionally, he was responsible for developing and conducting research in spatial disorientation and high acceleration (Gz forces) induced loss of consciousness (G-LOC). He was a science and engineering expert for the USAF, Air Force Research Laboratories and

Wyle Laboratories, Inc. on joint military (Air Force and Navy) G-LOC studies performing analysis of physiological data, i.e., Auditory Evoked Responses (AER), electroencephalograms (EEG), electrocardiograms (ECG), electro-oculograms (EOG), Oxygen saturation (SaO2), and Tracking Tasks Performance data.

Joseph P. Lessard, is the Vice President of Americas for Globeleq, Inc. Globeleq Inc. is a global owner and operator of power assets focused on the emerging markets. He is responsible for all business development activities in Latin America and the Caribbean. Mr. Lessard received a B.S. in Electrical Engineering in 1987 and an M.B.A. in 1994 from Texas A&M.

As an officer in the U.S. Navy, Mr. Lessard trained in nuclear power and served on the ballistic missile submarine USS Alabama.

Mr. Lessard entered the private power industry in 1994 as a project manager for the Coastal Power Company. The following year he was named Regional Managing Director with responsibility for business activities in Southeast Asia. In 1997, Mr. Lessard shifted his attention to the U.S. power market as Managing Director of the Northeast United States. He had profit and loss responsibility for three power plants and led a successful acquisition of a fourth power plant.

Mr. Lessard left Coastal Power Company in 1999 to form Hart Energy International, a start-up power company focused on the aggregation of power asset investments in Latin America. Hart Energy's successful direction of two acquisitions – EGE Haina in the Dominican Republic and Entergy's Latin American portfolio – led to the launch of Globeleq, Inc. in June 2002.

At Globeleq, Mr. Lessard has led the company's efforts in Latin America including the acquisition of a 200MW hydroelectric company, the divestiture of two non-strategic assets, the development of a greenfield thermal power plant, and the placement of two local bond issues. He is currently directing three greenfield development projects in Guatemala, Panama and Peru.